Mapping and Charting in Early Modern England and France

Toposophia: Sustainability, Dwelling, Design

Toposophia is a book series dedicated to the interdisciplinary and transdisciplnary study of place. Authors in the series attempt to engage a geographical turn in their research, emphasizing the spatial component, as well as the philsophical turn, raising questions both reflectively and critically.

Titles in the Series

Mysticism and Architecture: Wittgenstein and the Meanings of the Palais Stonborough, by Roger Paden
When France was King of Cartography: The Patronage and Prodution of Maps in Early Modern France, by Christine Marie Petto
Incapacitating Biochemical Weapons: Promise or Peril?, edited by Alan Pearson, Marie Chevrier, and Mark Wheelis
Environmental Dilemmas: Ethical Decision Making, by Robert Mugerauer and Lynne Manzo
The Timespace of Human Activity: On Performance, Society, and His-tory as Indeterminate Teleological Events, by Theodore R. Schatzki
Metamorphoses of the Zoo: Animal Encounter after Noah, edited by Ralph R. Acampora
Reading the Islamic City: Discursive Practices and Legal Judgment, by Akel Isma'il Kahera
Spoil Island: Reading the Makeshift Archipelago, by Charlie Hailey
Remembering Places: A Phenomenological Study of the Relationship between Memory and Place, by Janet Donohoe

Mapping and Charting in Early Modern England and France

Power, Patronage, and Production

Christine Marie Petto

LEXINGTON BOOKS
Lanham • Boulder • New York • London

Published by Lexington Books
An imprint of The Rowman & Littlefield Publishing Group, Inc.
4501 Forbes Boulevard, Suite 200, Lanham, Maryland 20706
www.rowman.com

Unit A, Whitacre Mews, 26-34 Stannary Street, London SE11 4AB

British Library Cataloguing in Publication Information Available

Library of Congress Cataloging-in-Publication Data Is Available

ISBN 978-0-7391-7536-1 (cloth : alk. paper) -- ISBN 978-0-7391-7537-8 (electronic)

∞™ The paper used in this publication meets the minimum requirements of American National Standard for Information Sciences Permanence of Paper for Printed Library Materials, ANSI/NISO Z39.48-1992.

Printed in the United States of America

Contents

List of Figures vii

Acknowledgments ix

Introduction xi

Abbreviations xxiii

1 Cartographic Imagery and Representations of Power 1

2 Mapping the Land: County and Regional Mapping in England and France 45

3 Chart Making in England and France, and Charting the English Channel 79

4 Paper Encroachments: Colonial Mapping Disputes in the Americas 123

5 Charting the Seas of the East Indies: Commercial Opportunism versus Royal Approbation 161

Conclusion 195

Bibliography 203

Index 215

About the Author 225

List of Figures

Fig. 1.1	Frontispiece from Saxton's *Atlas of the Counties of England and Wales* (1579).	7
Fig. 1.2	Frontispiece from Bouguereau's *Le Théatre Françoys* (1594).	12
Fig. 1.3	Detail from Vassalieu's *Portrait de la Ville . . . de Paris* (1609).	16
Fig. 1.4	Detail from Gomboust's *Lutetia, Paris* (1653).	20
Fig. 1.5	Norden's *London* (1593). Facsimile from the London Topographical Society, 1899.	23
Fig. 1.6	Frontispiece from Ogilby's *Britannia* (1675).	29
Fig. 1.7	Detail from Ogilby's London (1677/1682). Facsimile from the London Topographical Society, 1904.	30
Fig. 1.8	Detail from frontispiece of Jaillot's *Atlas Nouveau* (1684).	32
Fig. 1.9	Frontispiece from Jaillot's *Atlas Français* (1695).	34
Fig. 1.10	Detail from the cartouche of Nolin's *Le Royaume de France* (1693).	36
Fig. 1.11	Cartouche from Nolin's *Les Environs de Paris* (1698).	37
Fig. 4.1	Detail from Bellin's *Carte de l'Amerique Septentrionale* (1755).	149

Fig. 4.2 Mitchell's *Map of the British and French Dominions in North America* (1755). 150

Acknowledgments

I will open my list of grateful thanks with a rather non-academic statement, but most who have taken on the challenge to write monographs will nod their heads in agreement with its sentiment. I would like to begin by thanking every person who has listened to my 'tale of woe' (regarding this project), nodded politely, sometimes pretended they understood what I was babbling on about, and generally left me feeling a little bit better for having unloaded on them.

I begin again as I did for my first book to thank a few professors and mentors whose encouragement has made my research possible if not my entire teaching career. I thank Karl Petruso (now, Dean of the Honors College at the University of Texas at Arlington) who so many years ago at Boston University spoke kindly to a lost undergraduate and encouraged her to study in Greece. That experience opened up opportunities that led to my career in academia. Hopefully, Karl is not sorry that he added one more to the bunch. To the late Richard S. Westfall I again express gratitude for the gentle guidance, advice, and reassurance that with hard work graduate school could be survived. For the years of support in graduate school and in the early years of my career, I thank Helen Nader again for agreeing to direct my thesis and providing the opportunity to teach for a year at the University of Arizona. Finally, he may find this odd and may cringe at my writing, but I thank again Raymond Smith for having faith in my abilities and offering me the opportunity to develop my own writing and teaching skills while still in graduate school. Those experiences have proven essential to my career in the classroom. I may be a researcher, but I am also a teacher. Raymond helped me continue to foster my love of teaching and the ability to share my research with a wider audience.

Along the way I have been fortunate to receive several research grants from Southern Connecticut State University which allowed me to conduct my work in the archives in Paris and London. The current leadership at Southern is most supportive of faculty research from president Mary Papazian to the former provost, Marianne Kennedy, my former dean, DonnaJean Fredeen, current dean, Steven Breese, and colleague and associate dean, Bruce Kalk. I am fortunate to have an engaged audience who more than once nodded understandingly at my 'tale of woe.' Our library staff has been most helpful, in particular, Alba Reynaga and Shirley Cavanagh.

The support and enthusiasm for my project by the various associates and map librarians in Chicago, Paris, London, and New Haven has been most appreciated. Above all the early guidance by David Buisseret, James Akerman, Tony Campbell, Mireille Pastoureau, Monique Pelletier, Josef Konvitz, Mary Pedley, and Matthew Edney set me on the path to begin my research in French map making and then to the comparative work with English map making. Even brief words of advice coming at strategic times in the process were instrumental to the forward progress. I believe when I babbled on to them in my 'tale of woe,' they understood what I was talking about.

To Gary Backhaus who gave me the opportunity to publish my work and encouraged me to continue my research, as well as to consider co-editing a journal, I am most thankful. Robert Mugerauer and the anonymous reviewer have been most helpful in improving the manuscript. To Troy Paddock, Peter Nikola, and Alex Zukas who continue every year to comment on my conference presentations and encourage me to continue my work, I owe gratitude. Again to Troy Paddock, Nikos Chrissidis, and Vincent Pitts my colleagues who read (and re-read) the manuscript at various stages, I could not accomplish what I have without your support.

To my New Hampshire family that knows me less as the professor and more as the neighbor with trenches to dig, trees to cut down, and more often with her head in a book. To Joe and Anne, June, Mrs. Hoage, and Mary Sullivan who have not only fed me delightful food but continue to feed my spirit, thank you.

Finally, I began writing this book only a few months before the death of my beloved mother and truly best friend. With her passing the trinity of my life (my father, my mother, and my sister, Terry) was now gone. I was orphaned but fortunate to belong to a big family and owe the joy of my subsequent life to my friends and colleagues, but even more to my older brothers and sisters. On many occasions they have patiently listened to that 'tale of woe' and offered words of advice and encouragement. And so while I still so greatly miss my trinity, I dedicate this book to my brothers and sisters and their *better* halves. Marilyn and Frank, Ellie and Buddy, Al and Alleda, Dave and Ellen, John and Kathy, and Peter and Pam. You teased when I was young and continue to do so, but you are always there for me. Thank you!

Introduction

Many years ago I was offered the opportunity to be a research assistant to the historian of science and Newton scholar, Richard S. Westfall (d. 1996). While carrying out my investigations on the astronomers and mathematicians working at the Observatoire de Paris (Paris Observatory), I found myself drawn to what would become the national mapping project initiated by members of the Académie Royale des Sciences (Royal Academy of Sciences, hereafter Académie) as part of the program directed by Louis XIV's finance minister, Jean Baptiste Colbert (1619–1683), to serve the *gloire* of the Bourbon crown. Coming from the history of astronomy, I was comforted by the plan to draw from astronomical observations greater knowledge of the lay of the land and subsequently the shape of the earth. Armed with thoughts of geometric projections, instrumental precision, and skilled observers, I was ready to pronounce a judgment of sorts on the maps produced in Paris in the seventeenth and eighteenth centuries. Much to my surprise the trickle down of accurate latitude and longitude values did not revolutionize map printing though I noticed that maps were often advertised as drawn from the observations of the members of the Académie and the most up-to-date reports. It was in the process of examining hundreds of printed maps offered by Parisian map makers and map publishers that I decided to put aside the strictly technical aspects of map making to further examine their historical context, the motivations of production, and how the maps were received. This broader vision brought the work of the late Brian Harley into the sphere of my research and provided the motivation to study maps as instruments of power; an approach I adopted for my first book, *When France was King of Cartography: The Patronage and Production of Maps in Early Modern France* (2007).

I have looked to Harley's work once again for this current project, a comparative study of English and French cartography (and hydrography) in the early modern period, but in the interim I have tried to digest and embrace the works by those geographers and historians who have extended and critiqued the works of Harley. As I am interested first and foremost in the relationship between the state and the cartographic enterprise, I continue to investigate map making as a tool of power, but I do not make the general assumption that all maps must be instruments of power. There are three main threads of this study. First, I investigate the relationships that existed between powerful institutions (e.g., the Paris Observatory, the Académie, the English East India Company, the Board of Trade and Plantations, etc.) and social elites (e.g., monarchs, royal ministers, provincial officials, archbishops, nobles, gentry, etc.) on the one hand and the producers (engravers, geographers, surveyors, academicians, pilots, etc.) and publishers of atlases, maps, and charts from the late sixteenth century to the late eighteenth century on the other. How did the atlases, maps, and charts that emerged from these relationships influence a viewer's understanding of the world? As these relationships involved direct or indirect support for the map making enterprise, what was the role of patronage? Second, intertwined with these interactions were the practices and practitioners of the commercial map trade which provides greater depth to the map's social context and powerful associations. How influential was the trade in the creation of atlases, maps, and charts, and did it inhibit or enhance the connections between institutions/powerful elites and map makers? Finally, of course, there are the cultural artefacts themselves, that is, the atlases, maps, and charts produced during this roughly 250-year period. What role did these cartographic productions play in the world of early modern England and France? This work would not have been possible without the guidance from numerous monographs, articles, and discussion papers some of which I present below while others are addressed in the chapters that follow.

Even over twenty years after his death, all historians who engage themselves with the study of the role of maps, charts, and atlases in European history must pay respect to the works of Brian Harley.[1] Beginning at least in the mid-1980s, he urged historians to look at geographical works as socially constructed texts, that is, "to search for the social forces that have structured cartography and to locate the presence of power—and its effects—in all map knowledge."[2] The rise of state-directed cartography in early modern Europe allowed the state to gain power over knowledge.[3] Harley and Denis Wood argued that "cartography was primarily a form of political discourse concerned with the acquisition and maintenance of power."[4] Wood, however, wrote that "Harley was never able to conceive of the map as other than a representation of reality; was never able to grasp the map as discourse function; was never able to understand that the heart of the problem wasn't the way the map was *wielded* but the map function itself."[5] In an effort to move the discussion away from the map as a representation and to address "situational pragmatics" John Pickles using Wood writes, "Maps are made because of the needs of particular social situations; they are made to fulfill a particular function. . . . the map 'exists in its inscription.'"[6] Treating maps as inscriptions, Pickles focuses "on the ways in which mapping and the cartographic gaze have coded subjects and produced identities. . . . [what] work maps do, how they act to shape our understanding of the world, and how they code that world."[7] Rather than engaging with the debate over maps as mirrors of nature or ideologically-laden mirrors of nature, Pickles looks at maps as producers of nature. "Maps and mappings precede the territory they 'represent.' . . . Territories are produced by the overlaying of inscriptions we call mappings."[8] Moreover, "[r]ather than a determinate reading of the power of maps that seeks to uncover in a literal sense the authorial and ideological intent of a map (who made the map and for what purpose), Pickles expresses caution in fixing responsibility in such a manner, recognizing the multiple, institutional and contextual nature of mapping."[9] Such an approach not only provides a broader opportunity to investigate the discourses of maps, but also a more nuanced reading of the processes of map making and the subsequent cartographic productions.

These cartographic productions are also rhetorical texts, that is, persuasive documents that participated in the contemporary political, social, or scientific discourses. Rhetoric here is used in the classical sense rather than as a pejorative term. Harley wrote that map makers employed "the common devices of rhetoric such as invocations of authority (*especially* in 'scientific' maps) and appeals to a potential readership through the use of colors, decoration, typography, dedications, or written justifications of their method. Rhetoric may be concealed but it is always present, for there is no description without performance."[10] Decorative elements that are more easily accepted as persuasive were not to be viewed as "incidental"[11] or "insignificant marginalia."[12] "Decorative title pages, lettering, cartouches, vignettes, dedications, compass roses, and borders, all of which may incorporate motifs from the wider vocabulary of artistic expression, helped to strengthen and focus the political meanings of the maps on which they appeared."[13] In taking a phenomenological approach, Christian Jacob reminds the reader that ornamentation and figurative elements orient the viewer's gaze even creating "a dialectical movement between the sum and its parts."[14] He writes that the border (though I would add the other decorative elements as well) "influences the perception of the image by furnishing it with a space and a form."[15] These elements "guide the gaze through a series of stages, from the space outside the image to the space within it."[16] The patron, the map maker, the engraver, or even the map publisher may have played a role in the determination of the decorative elements which could be read by the map viewer as an amplification of territorial sovereignty, of an authoritative methodology, of local identity, or of mercantilist value. How did these significant elements enhance the map viewer's understanding of the work and his/her world?

In looking at maps as historical products and as contingent and relational, Matthew Edney outlines the cartographic modes that existed between 1500 and 1800 that historians often subsume under map making.[17] These distinctions and developments are important for my comparative study on English and French map making covering roughly the same period. The cartographic enterprise in the sixteenth century and into the late seventeenth century encompassed chorography, charting, and topography.[18] In England and France, chorography was small scale mapping of regions as well as the world, and its practitioners can best be described as practicing the exegetical method, that is, "the evaluation and reconciliation of geographic information of different types (textual, itinerary, other cartographic, pictorial) and of different dates (both ancient and contemporary) to produce a single map."[19] In France, adherents of this practice were called *géographes de cabinet* or geographers of the study. In England and France, charting over the course of this period borrowed mostly from the Dutch and still relied on written sailing directions, coastal views, and small manuscript charts. At the end of the seventeenth century both England and France would produce their own national marine atlases. Topography in this period began with the work of surveyors who created large scale maps (1:100,000 or larger) many of which remained in manuscript form for landed elites and the work of engineers who were sent by their monarchs to map areas of strategic importance. In the second half of the seventeenth century, the mode of geodesy emerged out of chorography and topography, and its representatives armed with the practice of making astronomical observations and mathematical measurements produced "abstract triangulation diagrams."[20] These were the foundational pieces for the work to establish the meridian of Paris and, with triangulation, to set up a network of triangles to aid in the creation of the national map of France, that is, the Cassini project. In England, triangulation would not become a common practice until the last decades of the eighteenth century.

As Edney suggests, by the early eighteenth century—earlier in France with the Cassini project—these modes merged into the "ambiguous" mode of mathematical cosmography. With improvements to and developments of mathematical tables, instrumentation, precision, and institutional or state attention, mapping in the Enlightenment period benefitted from an epistemology that joined empiricism, that is, the acquisition of knowledge through experience and experiment, with rationalism, that is, the acquisition of knowledge through the application of reason. "The cosmographer's habitual reconciliation of information from varying sources and of varying quality to produce a single map of undeniable worth, comprehensiveness, and utility has been recast as a metaphor for the fundamental belief of Enlightenment philosophy and the encyclopedic mentality, that rational debate and enquiry will lead ultimately to unified knowledge." The bureaucratic state would be well served by such maps, and its patronage of the enterprise provided it with "power, through the acquisition and control of geographical information; legitimation, through the willing submission of loyal servants; and prestige, through the visible support of the arts and sciences." In sum, "mapmaking was integral to the fiscal, political, and cultural hegemony of Europe's ruling elites."[21]

By the last decades of this century, France had embarked on a second national survey generated out of military/strategic concerns, and Britain followed suit with the Ordnance Survey tellingly housed in the Board of Ordnance. These mapping project were part of a new mode which Edney identifies as systematic mapping, created to serve the modern bureaucratic surveillance state "willing and able to support an expensive procedure in order to make more efficient the ability to marshal military resources and so wage war more effectively."[22] Such an important (and costly) endeavor embraced a "rhetoric of certainty," that is, "that the map replicates the territory's structure precisely and accurately."[23] This rhetoric, however, was presumed in the other modes as well. For example, "chorography, incorporated within mathematical cosmography via [Enlightenment] encyclopedism, is made to be directly equivalent to the instrumentalist-driven systematic survey. The uncertain art of the regional mapmaker is held to be comparable to the exact science of the surveyor." As a consequence without the acknowledgement of various cartographic modes in the march to greater precision, that is, the conception of linear progress and positivism, earlier maps or maps from a different mode fall victim to the "rhetoric of certainty." Historians, therefore, must not only contextualize the maps within the period of history, but also (and perhaps more importantly) recognize the "intellectual character of mapmaking as an intellectual, technological, social, and cultural process."[24]

PLAN OF THE WORK

This book is a comparative study of the production and role of maps, charts, and atlases in early modern England and France with a particular focus on Paris, the cartographic center of production from the late seventeenth century to the late eighteenth century, and London, which began to emerge (in the late eighteenth century) to eclipse the once favored Bourbon center. The themes that carry through the work address the role of government in map and chart making. In France, in particular, it is the importance of the centralized government and its support for geographic works through a broad and deep institutional infrastructure that James McClellan III and François Regourd studied in their work, *The Colonial Machine.*[25] While they have looked at these institutions as integral to the process of colonization, I also tap these institutions for their involvement in domestic affairs. In *Cartography in France, 1660–1848*, Josef Konvitz wrote of cartography's need for "consistent, visionary leadership . . . and a lot of money" which Louis XIV's government "began to provide."[26] In the second half of the seventeenth century, the Bourbon crown established and subsidized institutions such as the Académie and the Observatory with agendas to address cartographic projects. In the following century institutions such as the Hydrographic Office of the Navy and the Compagnie des Indes Orientale (East India Company) would support charting and mapping projects for the purposes of navigational safety, colonial expansion, commercial interests, and military strategy. In addition, the Crown supported groups such as the *géographes du roi* (royal geographers), *ingénieurs-géographes* (geographical engineers), and hydrographers directly through a pension or through a connection to a royal institution.[27] Such was the Machine that drove the cartographic and hydrographic enterprise.

Prior to the late eighteenth century in England, Harley suggested that there was no central controlling agency or institution for map, chart, or atlas production, and any official power was imposed through the market rather than through the establishment of institutions.[28] In her book, *The Commerce of Cartography*, Mary Sponberg Pedley wrote that there was no centralized support for the cartographic enterprise and that "the parliamentary government reduced the influence of even the most cartographically inclined monarch," and saw little value in fostering or supporting scholar-geographers or a national survey.[29] It remains to be seen, however, what informal, unofficial, or intermittent role was played by Whitehall, Parliament, the Royal Society, the Admiralty Office, the Board of Longitude, and the Board of Trade and Plantations among others. Furthermore, as patronage of the cartographic enterprise from social elites—the nobility and gentry often attached to the institutions above—did not involve their direct power in the trade, there appears to be no state or centralized influence in the market either.

Chapter 1 presents a study of the decorative embellishments found on English and French maps and in atlas frontispieces from the sixteenth to the seventeenth century. Numerous works have aided my investigations for this chapter, but the work of Peter Burke on the fabrication of the image of Louis XIV and Richard Helgerson's work on Elizabethan imagery have been most helpful.[30] The chapter begins by looking for the central image of sovereignty on (or in) the works and the historical context of these geographical works displaying symbols of dominion. These elements from the most entertaining representational layer of a map often reaffirmed territorial sovereignty for the patron of the cartographic work and emphasized his (or her) dominion to those who viewed the work. In England over time the central (or principal) position of the monarch who may have appeared in portrait or in image or symbolized by the royal coat of arms slipped from the central position to be joined (or replaced) by representations of the various social elites and powerful corporate bodies. In France influential imagery presented Bourbon rulers boldly assuming sovereignty over the land and even holding the central position in imagery while Europe, Asia, and America looked on in awe. In the eighteenth century the baroque sensibilities of art had passed and so had the day of the personal monarch. In imagery the bureaucratic state might still appear as armorial bearings but the rhetoric of power that had fueled the principal position of the sovereign was redirected to a scientific discourse that championed certain knowledge for the administration of the enlightened state.

Chapter 2 introduces the reader to the development of county and regional mapping in England and France from the seventeenth to the eighteenth century. I have already mentioned (and will express appreciation here for) the works of Pedley, Harley, Edney, and Konvitz, but at least two other works should be referenced: the well-known work on French ecclesiastical maps by François de Dainville as well as the much more recent work by Andrew Macnair and Tom Williamson on William Faden and county mapping.[31] The chapter begins with the epistemological shift that occurred in the second half of the seventeenth century which allowed those talented map makers to join the more traditional exegetical method with the empirical information from field observations and measurements. The educational background of many French map makers and the institutional infrastructure created to serve the Colonial Machine for interests abroad as well as the national mapping program set those map makers and France at the forefront of cartographic activity. The national mapping project led by succeeding members of the Cassini family motivated regional mapping from the diocesan level to the provincial level in efforts to serve political, economic, military, and religious concerns. Those projects that paralleled the Cassini efforts faced financial and organizational challenges. Despite the talented figures involved, it was clear that participation in the central project brought a greater chance for successful completion of the cartographic productions both for the state and the regional government. The visionary leadership and encouragement of mapping made cartography one of the standard operations of the modern bureaucratic state and created a professional cartographic core for those operations of the nineteenth century.[32]

Talented surveyors and mathematicians in England were disadvantaged by the attention to the marketplace rather than to the scientific discourse for their cartographic productions. In England, John Ogilby's attempted county atlas project in the second half of the seventeenth century remained incomplete as such an endeavor required support from the crown or county elites. Estate surveying out of which many efforts for county maps would emerge was simply sufficient enough for elites to view their properties. While they were willing to subscribe to a county map project, they were unwilling (or unable) to personally finance it. Many county map projects were not commercially feasible for map makers either who would need to create an infrastructure for the work from the survey to the drawing, to the engraving, and finally the printing. In the second half of the eighteenth century, projects for county maps were re-energized by the influence of the Society for the Encouragement of Arts, Manufactures and Commerce, but still the lack of centralized institutional support left a national effort in doubt. Established in the last decade of the eighteenth century, the Ordnance Survey and its team of trained surveyors and engineers and centralized production allowed England to finally realize a national mapping effort in the opening decades of the nineteenth century.

Chapter 3 takes the reader from the English county or French province to the English Channel and the respective home waters and coastlines of England and France. The work of Richard Harding on naval warfare provided the historical context for many of the hydrographic productions discussed in this chapter. In addition to the works of David Buisseret and Peter Barber on French and English mapping respectively, the masterful work by Olivier Chapuis on France's hydrographic office and Susanna Fisher's work on British chart making to name only a few were nearly constant companions.[33] The chapter begins with a presentation of English and French navigational works and chart making before 1700. Prior to the establishment of Paris as a cartographic (and hydrographic) center of production and the development of a map trade in London, both countries relied heavily on Dutch works. In the last decade of the seventeenth century after over one hundred years of attention to coastal fortifications and some coastal surveying, the crowns of both England and France initiated and supported (to varying degrees) their first maritime atlases. Collins' *Coasting Pilot* found a comfortable commercial niche and was continually published by Mount and Page into the second half of the eighteenth century with little or no updating. France's *Neptune François*, considered superior to the English work and even copied by the Dutch, had to be rescued by the clerk of the navy's hydrographic office in the middle of the eighteenth century for republication, a testament to the conservative nature of pilots. While Britain had continued to rely on commercial chart makers, the French crown (under a regency government) established a hydrographic office to produce charts for the royal vessels under the orders of the secretary (or minister) of the navy, another institution of the Colonial Machine. After introducing the role of chart making in Britain and France, the second half of the chapter focuses on the charting of the English Channel and the importance of institutional support to produce authoritative works within an infrastructure focused on improving charts for the surety of navigation rather than the vagaries of the marketplace. The Channel was a strategic waterway for military/naval operations as well as European and colonial commercial activity, and both governments endeavored to control it. Charts were a powerful tool in this contest.

Chapter 4 relates the entertaining story of English and French mapping of their colonial possessions in America and the controversy that such cartographic works created over issues of encroachment, boundaries, and identification of territory. Louis De Vorsey's work on the colony of Georgia, as well as Edney's work on John Mitchell's map, and Pedley's article on the cartographic dispute over Nova Scotia provided fuel for my study of the colonial encroachment clashes.[34] Much of the mapping for the Americas, in particular North America, embraced an effort to have geographical knowledge of place(s) but also a means of declaring territorial legitimacy in the face of British (or French) pretensions (depending on one's point of view). As Britain and France intended to solidify and expand their colonial interests in America, they both produced maps to define their holdings or challenge the definitions presented by one another. French map makers benefited from the institutional infrastructure of the Colonial Machine which often provided greater authority for their work. For example, Guillaume Delisle's map, *Louisiane* (1718), caused quite a stir as Britain perceived French territorial transgressions against their well-established colonial possessions. The response included not only audible complaints from the Board of Trade and Plantations, but also published English maps which aimed to challenge the French position but often had to rely on French cartographic sources to produce their own.

Several decades later, Britain and France found themselves once again in a dispute over the territory of Acadia (or Nova Scotia, depending on one's point of view). Despite a wariness on both sides concerning the use of maps to solve the dispute, their memorials included a discussion of evidentiary maps. While there was no official map of settlement in this case, English and French cartographers would produce maps and accompanying mémoires allied with their national position. An English audience would also see these debates played out in the pages of popular periodicals such as the *Gentleman's Magazine* or the *London Magazine* providing a forum to protest the pretensions of the perfidious French state. The story of these two disputes coupled with the understanding of the socially constructed nature of maps as well as the paradoxical rhetoric of the boundary disputes in which both sides decried the unreliability of maps and yet sought them out are the topic in this chapter.

Chapter 5 returns to nautical maps or maritime charts and takes the reader to the East Indies. While the works of P. J. Marshall and Philippe Haudrère provided general background information on the British and French East India Companies, the works of Andrew Cook addressed the navigational challenges of sailing to the East Indies, and Claude Briot and Howard Fry respectively provided information on the careers and works of two important hydrographers of the eighteenth century, Jean-Baptiste-Nicolas-Denis d'Après de Mannevillette and Alexander Dalrymple.[35] By the second half of the seventeenth century, England and France would come to challenge the Dutch who had superseded the Portuguese for trading dominance in the East Indies and China Seas. While Britain would be more commercially successful in challenging the Dutch and the French, the infrastructure of the Colonial Machine projected French chart making to a privileged position. By the mid-eighteenth century, d'Après had produced an atlas for the East Indies better than anything then available. In the absence of a better work in the commercial chart trade of London, the British would translate and copy this work for their own navigators. As discussed in Chapter Three, while Britain had continued to rely on commercial chart makers, the institutions of the French Colonial Machine supported more hydrographic works throughout the second half of the eighteenth century. In the last decades of this century as Britain's path to empire continued and its political role in India increased, Dalrymple's call for a hydrographic office was realized, first by the Company in 1779 and later by the Admiralty in 1795. By the end of the eighteenth century British dominance on the seas was certain and soon its ability to serve those pilots with Admiralty charts.

This work carries the reader from the second half of the sixteenth century in which both English and French map making and the map printing trade were only beginning to make their way in the field of cartography to the late eighteenth century at which time Britain's dominance was emerging and the French Bourbon State was on the verge of collapse. In map and chart production the center of power was moving from Paris to London. The maps and charts of the intervening two centuries (or more) have tales to tell and are the sources for the work that follows.

NOTES

1. See J. B. Harley, "Maps, Knowledge, and Power," in *The New Nature of Maps: Essays in the History of Cartography*, ed. Paul Laxton (Baltimore and London: Johns Hopkins University Press, 2001), 51–82; "Silences and Secrecy: The Hidden Agenda of Cartography in Early Modern Europe," in *The New Nature of Maps: Essays in the History of Cartography*, ed. Paul Laxton (Baltimore and London: Johns Hopkins University Press, 2001), 83–108; "Deconstructing the Map," in *The New Nature of Maps: Essays in the History of Cartography*, ed. Paul Laxton (Baltimore and London: Johns Hopkins University Press, 2001), 149–96.
2. Harley, "Silences," 86; and Harley, "Deconstructing," 152.
3. Harley, "Silences," 87.
4. Harley, "Silences," 85.
5. As quoted in John Pickles, *A History of Spaces: Cartographic Reason, Mapping and the Geo-Coded World* (London & New York: Routledge, 2004), 65.
6. Pickles, *Geo-Coded World*, 66.
7. Pickles, *Geo-Coded World*, 12.
8. Pickles, *Geo-Coded World*, 5.

9. Martin Dodge, Rob Kitchin, Chris Perkins, eds. *Rethinking Maps: New Frontiers in Cartographic Theory* (New York: Routledge, 2009), 13.
10. Harley, "Deconstructing," 163.
11. Harley, "Maps, Knowledge, and Power," 73.
12. Harley, "Deconstructing," 160.
13. Harley, "Maps, Knowledge, and Power," 73.
14. Christian Jacob, *The Sovereign Map: Theoretical Approaches in Cartography throughout History*, trans. Tom Conley, ed. Edward H. Dahl (Chicago: University of Chicago Press, 2006), 106.
15. Jacob, *Sovereign Map*, 107.
16. Jacob, *Sovereign Map*, 107.
17. Matthew H. Edney, "Cartography without 'Progress': Reinterpreting the Nature and Historical Development of Mapmaking," *Cartographica* 30, nos. 2 & 3 (1993): 54–68.
18. Edney, "Progress," 59.
19. Edney, "Progress," 60.
20. Edney, "Progress," 61.
21. Edney, "Progress," 62.
22. Edney, "Progress," 63.
23. Edney, "Progress," 64.
24. Edney, "Progress," 64.
25. See James E. McClellan III and François Regourd, *The Colonial Machine: French Science and Overseas Expansion in the Old Regime* (Turnhout: Brepols, 2011).
26. Josef Konvitz, *Cartography in France, 1660-1848: Science, Engineering, and Statecraft* (Chicago: University of Chicago Press, 1987), 4.
27. Mary Sponberg Pedley, *The Commerce of Cartography: Making and Marketing Maps in Eighteenth-Century France and England* (Chicago: University of Chicago Press, 2005), 26–34; and McClellan and Regourd, *Colonial Machine*, part I.
28. John Brian Harley, "Power and Legitimation in the English Geographical Atlases of the Eighteenth Century," in *Editing Early and Historical Atlases*, ed. Joan Winearls (Toronto: University of Toronto Press, 1995), 164.
29. Pedley, *Commerce of Cartography*, 199.
30. Peter Burke, *The Fabrication of Louis XIV* (New Haven, CT: Yale University, 1992); Richard Helgerson, *Forms of Nationhood* (Chicago: University of Chicago Press, 1992).
31. François de Dainville, *Cartes Anciennes de l'Eglise de France* (Paris: Librairie Philosophique J. Vrin, 1956); and Andrew Macnair and Tom Williamson, *William Faden and Norfolk's 18th-Century Landscape* (Oxford: Oxbow Books, 2010).
32. See Konvitz, *Cartography in France*, 22, 31.
33. Richard Harding, *Seapower and Naval Warfare, 1650–1830* (London: University College, London Press, 1999); David Buisseret, "Monarchs, Ministers, and Maps in France before the Accession of Louis XIV," in *Monarchs, Ministers, and Maps: The Emergence of Cartography as a Tool of Government in Early Modern Europe*, ed. David Buisseret (Chicago & London: University of Chicago Press, 1992), 99–123; Peter Barber, "England I: Pageantry, Defense, and Government: Maps at Court to 1550," in *Monarchs, Ministers, and Maps: The Emergence of Cartography as a Tool of Government in Early Modern Europe*, ed. David Buisseret (Chicago: University of Chicago Press, 1992), 26–56; Olivier Chapuis, *A la mer comme au ciel* (Paris: Presses de l'Université de Paris-Sorbonne, 1999); and Susanna Fisher, *The Makers of the Blueback Charts: A History of Imray Laurie Norie & Wilson Ltd* (Ithaca: Regatta Press, 2001).
34. Louis De Vorsey Jr., "Maps in Colonial Promotion: James Edward Oglethorpe's Use of Maps in 'Selling' the Georgia Scheme," *Imago Mundi* 38 (1986): 35–45; Mary Sponberg Pedley, "Maps Wars: The Role of Maps in the Nova Scotia/Acadia Boundary Dispute of 1750," *Imago Mundi* 50 (1998): 96–104; Matthew Edney, "John Mitchell's Map of North America (1755): A Study of the Use and Publication of Official Maps in Eighteenth-Century Britain," *Imago Mundi* 60, no. 1 (2008): 63–85.
35. P. J. Marshall, "The English in Asia to 1700," in *The Oxford History of the British Empire: The Origins of Empire,* ed. Nicholas Canny (Oxford: Oxford University Press, 1998),

264–285; P. J. Marshall, "The British in Asia: Trade to Dominion, 1700–1765," in *The Oxford History of the British Empire: The Eighteenth Century*, ed. P.J. Marshall (Oxford: Oxford University Press, 1998), 487–507; Philippe Haudrère, *La Compagnie française des Indes au XVIIIe siècle*, tome I, 2nd edition (Paris: Les Indes Savantes, 2005); Philippe Haudrère, "The 'Compagnie des Indes' and Maritime Matters (c. 1725–1770)," in *Ships, Sailors and Spices: East India Companies and Their Shipping in the 16th, 17th, and 18th Centuries*, ed. Jaap R. Bruijn and Femme S. Gaastra (Amsterdam: NEHA, 1993), 81–97; Howard T. Fry, *Alexander Dalrymple (1737–1808) and the Expansion of British Trade* (Toronto: University of Toronto Press, 1970); Andrew S. Cook, "Establishing the Sea Routes to India and China: Stages in the Development of Hydrographical Knowledge," in *The Worlds of the East India Company*, ed. H. V. Bowen, Margarette Lincoln and Nigel Rigby (Rochester, NY: D. S. Brewer, 2002), 119–36; Andrew S. Cook, "Surveying the Seas: Establishing the Sea Routes to the East Indies," in *Cartographies of Travel and Navigation*, ed. James R. Akerman (Chicago: University of Chicago Press, 2006), 69–93; Claude Briot, "L'Apport de d'Après de Mannevillette dans les progrès de la Navigation en Mers des Indes au XVIIIe siècle," in *Autour de D'Après de Mannevillette: Savant Navigateur Havrais au siècle des Lumières*, ed., Eric Saunier (Havre: Imprimerie la Petite Presse le Havre, 2008), 33–60.

Abbreviations

AN — Archives Nationale, Paris
AR — Archives de l'Académie des Sciences, Paris
BL — British Library, London
BN — Bibliothèque Nationale, Paris
NL — Newberry Library, Chicago

Chapter One

Cartographic Imagery and Representations of Power

A friend once remarked that the decorative elements of a map are the most entertaining part to look at. I think all the parts are entertaining, but he did raise the issue that there are parts to a map; one might suggest that there are different representational planes or signifying elements in (or on) a map.[1] The map or graphical image is a representation of the land, a mediation of space between the landscape and the map viewer; within the constraints of scale, available sources, instrumental precision, and the skill of the map maker, it aims to be a mimetic representation. Layer upon this geometric framework a second one with the signifying elements that became over the years a normative set of key characteristics, that is, rivers, roads, secular and religious buildings, elevations, bodies of water, forests, and so on. Third, add to these topographical features place names of varying textual size and the deictic function of the map is complete. This is France; this is London; or even this is Normandy. To assist or amplify the deictic function, the map sheet includes yet a fourth layer of text which is usually set apart from the geometric framework and signifying elements, perhaps in the borders of the map sheet or in, say, a cartouche of variable size, style, and complexity of design. This layer—the most entertaining part—was the realm of rhetoric or persuasive iconography, scenery, or verbiage. One might recognize in this layer, decorative embellishments that emphasize or reiterate in imagery the sovereignty of the land, bold border delineations, or the religious, political, military, or socioeconomic function of the land. This representational plane may not only have been the design and work of an individual distinct from the surveyor, geographer, map compiler, or line engraver, but also may have carried a story all on its own in the ornamental cartouche or the series of decorative vignettes encircling the map in a framework contributing to a rhetorical message. These design elements often made the work more attractive and valuable but could increase the cost of the map production by 25 to 65 percent.[2] The decision to embellish, then, could not be taken lightly.

The purpose of this chapter is to investigate this entertaining representational plane on English and French maps and atlas frontispieces from the sixteenth and seventeenth century while keeping in mind its relational bonds to the map (or atlas) as a whole. The focus is on the imagery of power displayed in and on these geographical artifacts and what that imagery conveyed to the map (or atlas) viewer. The personal monarch was central to the imagery of dominion in the first late sixteenth-century English and French atlases, but political and economic developments shifted and/or intensified this centralization of power. English map imagery and production reflected not only decentralized and commercially driven trade but also a changing political landscape in which the role of the monarch was displaced from the center and forced or negotiated—it is unlikely that Charles I would have used the term negotiated—into a shared position of power. Prior to the Glorious Revolution the position of the monarch *on* the map may have reflected his relationship with Parliament. As the personal power of the monarch slipped from a central position on the map it was often joined by a myriad of symbols (e.g., coats of arms) of competing or complementary groups, that is, social elites and institutions. In seventeenth-century France with the rise of Louis Dieudonné (the God given) and future Sun King, however, map makers and map publishers operated within a much more centrally controlled society and often with the support of the crown. They were certainly interested in producing commercially successful works, and so they looked to the interests of their monarch. Such highly decorated geographic productions therefore served the glorification of the power of the Bourbon crown or specifically, Louis XIV. Such embellishments of power were evident even at the end of the seventeenth century when the wars of Louis XIV produced little to glorify. While this monarchical power came to support cartographic enterprises and made Paris the center for European map trade, the less directed (and less supported) London trade came to highlight not the crown but pragmatism and merchant activity. Here the Stuart flourishing kingdom was not by military arms as in Bourbon France but by commerce.

Though never ideologically neutral, these decorative elements *appear* the most intellectually approachable by any map viewer. I write *appear* as the modern map viewer may certainly recognize and delight in the various symbols, animals, mythical creatures, flora and fauna, and both female and male figures, but their representational use (or choice in the design elements) by the map maker, engraver, publisher, or patron may be quite opaque. In analyzing the compelling imagery on maps and atlas frontispieces, it is, therefore, important not only to contextualize the piece under investigation, but also to be cognizant that it is a socially constructed artifact that often participated in a larger cultural multiple media program.

TUDOR ICONOGRAPHY AND MAPPING THE GRANDEUR OF ELIZABETHAN ENGLAND

Among the more enigmatic examples in the Tudor multiple media program are the many portraits of Elizabeth I. For example, *The Rainbow Portrait* (c. 1600) has occasioned much analysis as a youthful Elizabeth (d. 1603) despite her more advanced age at the time of the work appeared grasping a rainbow with the phrase "Non Sine Sole Iris," or "No rainbow without the sun" written just above her hand.[3] Marcus Gheeraerts, a Dutch Protestant émigré to whom this work is attributed, adorned her with a mantle decorated with eyes and ears, and upon her dress sleeve was a jeweled serpent with a heart-shaped ruby on a golden chain hanging from its mouth and above its head an armillary sphere. She was draped with pearls and upon her head a piece with more pearls and a crescent moon. The work was meant to speak to the viewer of her prowess as a monarch. The rainbow was symbolic of peace; the serpent of wisdom; the armillary (or celestial) sphere of omniscience; the heart-shaped ruby of counsel; the pearls of chasteness/virginity; and the eyes and ears symbolic of fame (or renown) or alternatively, again, her omniscience drawn from her official information network.[4] This is an astounding and bewildering work of art. It is descriptive of her reign and a representation of her beneficent sovereignty as there would have been no peace, symbolized by the rainbow, without Elizabeth, the sun. She is well-informed by her counselors and royal officials, and while she seeks their advice, it is her wisdom that is highlighted. "Many things she sees and hears through them [i.e., her servants], but the Judgement and Election are her own."[5] Elizabeth is the sun that emerges from the storm clouds to bestow the rainbow.

The power and dominion of this Tudor monarch is more clearly and less enigmatically displayed in another work by Gheeraerts, *The Ditchley Portrait* (1592), commissioned by Sir Henry Lee (1533–1611), Elizabeth's champion from 1559–1590.[6] In this work Elizabeth stands on a Saxtonesque map of England and Wales, specifically on Ditchley in Oxfordshire, the location of Lee's family home. Lee intended to commemorate the honor of her royal visit, but the imagery and the text displayed on the canvas speak to a rhetoric of territorial dominion and the role of protectoress. Elizabeth is imbued with cosmic powers as in *The Rainbow Portrait*, but here she acts prior to the appearance of the rainbow as she banishes the storm clouds and ushers in the sun to shine upon her beloved kingdom. Painted a few years earlier, *The Armada Portrait* (1588) by George Gower similarly shows the storm clouds behind the Queen as Spanish ships wreck in stormy seas and the sun is before Elizabeth in sight of her victorious fleet.[7] In place of the flat map is the globe on which she confidently places her right hand perhaps on the North American continent while just above on a table to her right is the imperial crown. One is reminded of the works of John Dee and Richard Hakluyt (among others) who wrote of England's imperial destiny (see below). Such imagery also reminds one of the depths of meaning that may be present in that representational frame of a map in which the realm of rhetoric or persuasive iconography is found. In an effort to illuminate the fullness of the map, historians of cartography seek to resurrect the context of the work and to disassemble its constructed social space.

Nearly a century before Louis XIV's department of glory (see below) succeeded in commissioning and producing multiple media projects to contribute to his *gloire* and to construct the cult of the Sun King, Tudor and, to less success, Stuart monarchs and their champions also adopted, manipulated and circulated a program of imperial dominion manifested not only in period art and literature, but also in atlas frontispieces and dedications as well as map cartouches.[8] Such representations reminded and reinforced "the power of the sovereign before the people."[9] Simply put, Elizabeth was the manifestation of England. These representations of authority were, according to Kevin Sharpe, aimed at "a public which was perceived to be able to understand what they signified."[10] In other words, Elizabeth's mantle with eyes and ears was a calculated *fashion* statement.

Images of power and dominion grew out of responses to the political, economic, religious, and military actions in the second half of the sixteenth century. The cult of Elizabeth, according to Roy Strong, began to emerge, particularly, in the late 1560s as exiled Protestants (like Gheeraerts) viewed her as their savior against the hostility and ferocity of Catholic Spain.[11] Elizabeth was, however, also quite clear in her grasp of the role of the monarchy in England with respect to Parliament and to her (other) subjects. Politically, while she recognized the regional, local, and corporate interests, she was a staunch believer in the power of the monarchy as the unifying force of the kingdom. She clearly expressed her beliefs in her speeches of 1566 to Parliament regarding among other things the issue of marriage and succession. In early November she chided those assembled: "A strange thing that the foot should direct the head in so weighty a cause [i.e., succession]" and, a little further on, more strongly, "it is monstrous that the feet should direct the head."[12] As late as 1598 she was still reminding members of Parliament of their role as councilors only, "no king fit for his state will suffer such absurdities."[13] In the 1570s, with religious conflict still raging in the Netherlands, Elizabeth launched a military campaign to assist the Dutch Protestants revolting in the ten Northern provinces. With Elizabeth at the helm, the government moved more decisively in its imperial ambitions not only heightening the imagery and rhetoric of England's imperial destiny, but also exacerbating earlier fears of invasion requiring an assessment of England's readiness to cope—an assessment that could come in the form of a map.[14] The cult of Elizabeth evolved to display symbols of imperial power as well as mythologized forces of nature that solidified her dominion over the land.

Englishmen had recognized the usefulness of maps as tools of government beginning in the reign of Henry VIII as his Tudor government called for surveying efforts along England's coastline.[15] Historians of geography have argued that a map consciousness arose in England in the second half of the sixteenth century. According to Lesley Cormack, the increase in the number of those being educated in geography at Oxford and Cambridge helped to promote this map consciousness.[16] William Cecil, Lord Burghley (1571) and Elizabeth's secretary of state, well appreciated the usefulness of maps both to visualize the kingdom and to assess its resources.[17] Moreover, map savvy members of Elizabeth's Privy Council (e.g., William Cecil, Sir Francis Walsingham, and Christopher Hatton) had foreign examples to imitate such as Peter Apian's survey of Bavaria—the first regional survey—(1568) for Albrecht V, and the first world atlas, Abraham Ortelius's *Theatrum Orbis Terrarum* (1570) dedicated to none other than Philip II of Spain (and the Netherlands) and England's rival. Coupled with the more recent interest in Old English studies or antiquarianism there was a growing curiosity to investigate the interior regions of the country.[18]

In 1579 after previous attempts by others, Christopher Saxton (c. 1540–1610) produced his *Atlas of the Counties of England and Wales*. It included thirty-five maps—thirty-four regional or county maps covering fifty-two counties and a general map of England and Wales—all of which carried the royal arms, the arms of Thomas Seckford, Master of the Queen's Court of Requests and future Surveyor of the Court of Wards, and the identification of Saxton as the author. For sixteenth-century atlas frontispieces, there is simply nothing more stunning than that from Christopher Saxton's county atlas (see figure 1.1). As a participant in the multiple media program, the frontispiece of this work, among the early emblematic English title pages, visually reads as a declaration of authority, since between the covers of this atlas lay the depiction of Elizabeth's dominion.[19] She knew well the portrait that would make up the frontispiece, as she instructed Saxton to alter the image so that the queen's skirts would fall "more flatteringly."[20] At the center of this architectural title page, by the Flemish engraver Remigius Hogenberg, was the enthroned Elizabeth with scepter and orb (and flattering skirt) guarded by miniature statues at the top of her throne of the goddess Pallas Minerva (or Pallas Athena) with spear and shield while figures of the royal lion with shield graced the arms of the throne. Framing Elizabeth were Corinthian columns and a frieze below their bases of the goddess Diana (or Artemis) embracing a column and Venus (or Aphrodite) both classical images associating Elizabeth with the might of a warrior and the beauty of a goddess. Surmounting the structure was the royal coat of arms flanked by the crowned lion and the dragon while two Victories with laurel wreaths fill out the horizontal space above the frieze where just below the royal arms an oval image depicted Righteousness (with sword) and Peace (with olive branch) embracing as Justice looked on from the background. Most dominant after the central image of the queen were the dual figures of Geography with compass and globe and Astronomy with an armillary sphere just beside the columns at Elizabeth's side in active service to her majesty as the interior maps would show. Beneath the feet of Elizabeth were two cartouches with inscriptions; the first praised this "gentle governess," and the second acknowledged not only her wisdom, piety, and justice but also the stability of her reign that made her dear at home and celebrated abroad.[21] By this atlas, her dominion at home was celebrated not only *on* England and Wales, but across the counties all of which displayed her armorial bearings. To the left of these inscriptions a figure drew a map and to the right another gazed upon the heavens, but visually upon the grandeur of the enthroned queen, who had by 1579 ruled for over two decades and had begun to challenge Spanish hegemony to build an English empire. Influential scholars such as John Dee, Richard Hakluyt, John Wolfe and voyagers such as Francis Drake, John Hawkins, and Walter Raleigh called for England to take its "rightful place" at the helm as a seafaring nation and to participate in creating England's imperial identity. After all, Gloriana had already brought righteousness, peace, and justice to her realm.[22]

In the maps of her realm that followed, her presence on the map was denoted by the royal coat of arms or her portrait. For example, Saxton's twenty-sheet wall map of England and Wales (1583) included a fretwork cartouche with Elizabeth's coat of arms topped with the imperial crown and the supporters—the crowned lion and unicorn—along the side. Just below was a brief history of England from Caesar to Elizabeth as well as Seckford's presentation of the work to the Queen.[23] In 1592, after more than three decades on the throne and in the glorious aftermath of the defeat of the Spanish Armada, Jodocus Hondius, a Dutch engraver living in London at the time produced a map of England, Wales, and Ireland accompanied by a genealogical table on the map of the kings of England since the Norman Conquest. The rare second state of this map replaced the badge of the garter with a portrait of Elizabeth at the center of the genealogical table.[24]

Two years later, an anonymous author copied from Hondius's work and included the portrait of Elizabeth at the center of a genealogical table which extended from Dunkirk on the coast of French Flanders along the eastern coastline of England and Scotland reminiscent of the place of the unsuccessful and hazardous journey of the defeated Spanish fleet. In the portrait, Elizabeth is viewed seated with the scepter in her lap, the orb in one hand while

Figure 1.1. Frontispiece from Saxton's *Atlas of the Counties of England and Wales* (1579). *Source:* Courtesy of the British Museum, London

her other hand holds a branch of the tree on which two white roses blossom and lower on the branch, two red roses blossom, representative of the Tudor roses and an allusion to her lineage from both the Lancasterian and Yorkist families. Printed around the outer frame of the portrait is her title, Elizabeth, by the grace of God, Queen of England, France, and Ireland. To the upper right beside the table is the royal coat of arms topped by the imperial crown, supported by the crowned lion and unicorn, and encircled with the motto of the Order of the Garter: *Honi soit qui mal y pense*, evil to him who thinks evil.[25]

Less than a decade later in the year of Elizabeth's death, 1603, Hans Woutneel, a Flemish bookseller living in London, produced a similar work but now with the armorial bearings of the new sovereign.[26] Woutneel dedicated the map, *A Descripsion [sic] of the Kingdoms of England, Scotland, & Ireland*, to James I (James VI of Scotland) and also adorned the map with the genealogical tree of the kings of England, but rather than a central portrait of James the entire tree includes small portraits of all the ruling monarchs with James and his wife, Anne of Denmark, Queen consort, positioned at the top of the tree. Visually, the portrait of James does not dominate the space of the map and even more revealing is the depiction (and brief description) on the map of the crowning achievement of Elizabeth's reign, the defeat of the Spanish Armada. The monarch is still *on* the map, but the imagery of past success and past rulers denotes a sort of review for the new sovereign that was on the threshold of his rule, but England had changed. James would not sit enthroned on the frontispiece of a national atlas in recognition of his dominion over the land, and aside from some small private commissions for surveying work (see below), he would not fund such endeavors either.

Richard Helgerson has suggested that with the advent of Saxton's atlas containing distinct maps of regions and/or counties, an English audience was disabused of a dynastic primacy. Camden's *Britannia* (first edition to include maps, 1607), Drayton's *Poly-Olbion* (1612), and Speed's *Theater* (1611) among others, "contributed to a shift of primary loyalty from king to country, from the dynastic state to the land-based nation."[27] While Saxton's frontispiece revealed an enthroned Elizabeth, Camden and Drayton removed the queen; "The land has, in effect, displaced the sovereign and assumed for itself the signs of rule."[28] In Camden's *Britannia*, the royal coat of arms that were prominent on maps in Saxton's atlas are nearly gone with only eleven of Camden's fifty-six maps displaying the royal arms.[29] While many (thirty-six of forty-two) of Speed's county maps included the royal coat of arms, Helgerson notes that the armorial bearings are much reduced in size and joined by a host of other coats of arms, town plans, battle scenes, sights of interest, and the like, such as "features that direct attention away from the king and toward the country."[30] The numerous arms of the local gentry displayed on Speed's county maps "contributed to the emergence of a conception of England defined not exclusively in relation to its monarch, but to all the leading families of the gentry."[31] In seventeenth-century England, support for cartographic activity would not come from the crown but from private initiatives as well as the expansion and diversification of the London printing trade which produced maps and atlases for the purchasing tastes of the gentry and the growing middling sort. These elites might patronize particular works, but played no directing role in either map or atlas production.[32]

ORIGINS OF GEOGRAPHY IN THE SERVICE TO THE BOURBON CROWN

Across the Channel in sixteenth-century France where attentions were preoccupied with the Wars of Religion, there was no similar Saxton-type surveying project completed. David Buisseret described the French map consciousness of the early sixteenth century as "rudimentary," but there were talented geographers and mathematicians serving what few royal interests there were.[33] Oronce Finé, for example, produced maps for François I and was so favored to be appointed royal professor at the newly founded College Royale.[34] Buisseret has also written of Nicolas de Nicolay's incomplete project (c. 1560)—"a general account of France" with maps—for Catherine de Medici (1519–1589), but only a few manuscript maps are extant, as a larger atlas project never came to fruition due to the upheavals of the wars.[35] According to Hilary Ballon "[u]nlike the other monarchs of sixteenth-century Europe, the Valois kings displayed little interest in mapmaking, whether for decorative, military, or political purposes."[36] The government's more sustained interest in map making and an iconography linked with geography would not emerge until the seventeenth century, but there is a hint of such imagery by the Florentine widow of Henri II, Catherine de Medici, who had on display in her Parisian salon twenty-four large maps covering Europe and the world.[37]

In one royal ceremony, a goddess—representing Catherine de Medici—crowned and in antique dress was depicted holding a map of France above her head while at her feet rested a crane, a hare, and a dolphin all of which symbolized the vigilance, virtue, and readiness that the queen mother had taken over the previous ten years to sustain the kingdom.[38] This procession that accompanied the coronation ceremony (March 1571) of Elisabeth, Archduchess of Austria and wife of Charles IX, not only celebrated the royal couple, but acknowledged the dominant role of the queen mother in power politics. As the procession wound its way through the streets of Paris, participants, onlookers, and readers of the text published (in 1572) to mark the event would come upon the Fontaine du Ponceau (Second Arrondissement) adorned with this map-holding goddess. She was flanked by two small pillars, one with a scepter decorated with an eye and ear (reminiscent of the mantle of Elizabeth I) and participating in the message of vigilance and readiness, and the other with a broken sword beneath which was a goblet, a lute, joined hands, and two hearts providing a message of diligence in preventing war, promotion of confederation, and reconciliation of the opposing parties. Near the base of the fountain were seated four classical female figures, Lucretia, Queen Artemisia, Camilla, and Cloelia, representative of Catherine's chastity, piety, bravery, and service to her people. The authors of the text describe the map of Gallia as full of towns, cities, and villages near forests, rivers, mountains, and valleys. It was the queen mother, former regent and arguably the power behind the throne that sustained that vibrant, pacific, and bucolic image of France held with great care in the hands of this most gracious lady.[39]

In early 1589 Catherine de Medici died and over the course of the next several months the Valois King Henri III, so busy trying to keep hold of his kingdom and ward off the Guise family, lost at the game of intrigue and power politics. He was assassinated on 1 August 1589 after joining forces with Henri of Navarre to lay siege to the city occupied by the Guise and their Spanish allies. In these final years of the Wars of Religion, the Protestant, Henri of Navarre found himself heir to a throne that he would have to fight to achieve. Paris had risen in revolt (1588) after the assassinations of Henri de Guise, and Louis, cardinal de Guise, the leaders of the League. A Protestant king was, however, inconceivable, and the civil war that followed only continued to inflict hardships and horrors across France. To achieve the throne, this Protestant tyrant would have to become a Catholic savior for the French people.

Navarre, his courtiers, and a royalist parlement had lived in exile in Tours as Paris was occupied by the Catholic League and its Spanish allies. Maurice Bouguereau, editor and bookseller, worked in Tours with several others of the printing industry who had fled the city as well. On 12 June 1591, Bouguereau was awarded by the exiled crown a privilege of ten years to collect, correct, and augment maps and have them engraved to cover the lands loyal to Henri and those lands under his control.[40] There was no survey as part of this royal privilege.[41] Although Bouguereau's *Le Théatre Françoys* borrowed much from the atlases of Ortelius and Mercator and a few earlier French maps, he intended it to be recognized not only for its finely engraved maps (by the Flemish Protestant émigré, Gabriel Tavernier)—only one of which carried the royal coat of arms, but also as a political manifesto for Henri.[42] In fact, he condemned the League for fomenting rebellion against a rightful king in the guise of religion in order to hide its more sinister plan to seize control which would make France a puppet of Spain.[43] During the reign of this Bourbon king, mapping would emerge as a more permanent tool of the state, but first Henri had to obtain his kingdom.

He commenced his journey to Paris and the throne by abjuring Protestantism in July 1593, and after being crowned Henri IV in February, Paris opened its doors to him in March 1594. With the rapid return to Paris of printers, engravers, etc., and though incomplete, Bouguereau rushed his great work to print in Tours.[44] The architectural frontispiece is a triumphant arch similar to those created for the festive events of the earlier Valois celebrations and those that would follow this Bourbon king (see figure 1.2). The light comes from the east with the rising sun and upon the top of the arch two nude figures each suspend a cornucopia-like display symbolic of the prosperity that the reign of this ruler would usher in. At the center top between them is the crowned double emblem for France and the Navarre of Henri de Bourbon (Henri IV)—the fleur-de-lis and the cross and saltire chain (on a red ground with an emerald at the center). Beneath the double coat of arms are two crossed olive branches with a crowned H at the center, and surrounding this crowned heraldic piece are the two collars for the Order of Saint Michael and the Order of the Holy Spirit. Now crowned, Henri brings peace and religious harmony to the kingdom. In the opening of the arch is the title of the atlas "to the king" and to the left and right are the personifications of Geometry holding a ruler and a square in her left hand and dividers in her right while Geography holds a quadrant above her head and a map in her right hand. This Bourbon king whose military feats and great clemency had brought peace, prosperity, and harmony to the French people could see within this atlas, the work of geometry and geography, the just measure of the lands he held by his greatness.

On the following page, Henri appears in portrait from the waist up in armor with "Henry IIII Roy de France et De Navarre" printed around the oval frame. To the right of the frame is the personification of Navarre with her coat of arms and dressed in a garment adorned with the armorial symbol. She holds up a crown as if to crown the portrait of Henri while the figure to the left, France, holds up her coat of arms, the fleur-de-lis, just parallel to the crown in Navarre's hand. France also carries in her left hand the scepter entwined with an olive branch and a crown at the top. Below the portrait is a quatrain: "Here is pious Henry conqueror of the Spanish/ who holds two powerful scepters with a strong hand/ It is this king whose face bears the sign of glory/ Clemency in his soul and prowess in his heart."[45] Beneath the quatrain is a small map of France with the united coats of arms of Henri de Bourbon crowned and encircled with the collar of the Order of St. Michael in the lower left corner. This small map is similar to the larger map (1593) of Gallia (France)—the only map in the atlas to carry the double coat of arms of Henri IV—and taken from the work of Petro Plancio. In the upper right corner was a dominant double coat of arms, the collars of the Order of the Holy Spirit and the Order of St. Michael, and implements of war as well as symbols of peace and plenty.[46] In the same year a similar map by Plancio

Figure 1.2. Frontispiece from Bouguereau's *Le Théatre Françoys* (1594).
***Source:* Courtesy of the Bibliothèque Nationale, Paris**

(perhaps from the same plate) was published in Paris by Paul de la Houve. This map included a portrait of Marie de Medici and her coat of arms in the upper right corner of the map and a portrait of Henri IV and the double coat of arms in the left corner with a brief text reflecting the benefit of Navarre's peace to all of Europe.[47] This imagery coupled with the quatrain and the sonnets found in the opening pages of Bouguereau's atlas paints a portrait of a valiant king. All of France (perhaps even Europe) sang the praises of this king to whom Mars has ceded all honors in war and just as Hercules successfully defeated the Hydra, good king Henri brought down the League and returned the kingdom to peace and prosperity. It was only fitting that Geography rendered a drawing of this fortunate kingdom for this very Christian king. Geography, then, was in service to the crown and by extension, the well-being of the state.

FRAMING THE ROYAL CITY OF PARIS

Fifteen years after Bouguereau's atlas in 1604, geography was still in service to this king, when Benedit de Vassallieu called Nicolay, royal engineer, surveyed the city "anew" rather than relying on earlier maps of the city as models.[48] This map (approximately 3 feet x 2.5 feet), a bird's eye view of the city, coupled with the aesthetically pleasing iconography from the skilled hands of Léonard Gaultier (1561–1641) provided a visually attractive and no less powerful reading for Henri IV. Jean Le Clerc (1560–1621/22) the publisher with whom Gaultier was known to work and also an associate of Bouguereau had fled to Tours with many others of his trade during the upheaval of the last years of the Wars of Religion. Now returned to Paris and a great supporter for the Bourbon king, Le Clerc published Nicolay's map which included a framed image of the king, "an element never before seen on a plan of Paris."[49]

The imagery from Bouguereau's atlas is repeated on this map of Paris and even without the numerous sonnets and quatrains, the message was still the same: Henri de Bourbon brought peace to the kingdom. The map of Paris, however, has special significance as the city had held out against its king. In 1589, as Henri of Navarre became nominally the king of France, the Catholic League and its ally, Spain, occupied Paris and had no intention of allowing France to fall into the hands of a Protestant. To assume the throne, Henri had to weaken and ultimately defeat the League and its Spanish allies, take the city of Paris, and take the crown. Henri laid siege to the city of Paris at least three times, and while the people of Paris were not necessarily happy under Guise control, they could not endure the thought of a Protestant king, and their gates remained closed.

From 1589 to 1594 Paris government officials had tried to remain independent dodging the Catholic League and the worrisome pretensions of their allies, the Spanish, and holding out against the possibility of a Protestant king. Once the king was anointed and crowned in Chartres, Paris, the royal city, had to bow in supplication. On 22 March 1594, Henri with part of his army approached the entry to Paris at the farthest corner where the Louvre and the Tuilleries run along the Seine called Porte Neuve (New Gate).[50] On Nicolay's map of Paris, this site appears in the lower right corner just above the left tip of the city's coat of arms allowing the map viewer's gaze to move diagonally upwards approximately following Henri's processional route to Notre Dame and to the royal coat of arms in the upper left corner of the map.

On that day in March 1594 at Porte Neuve, Henri was met by the governor of Paris (Brissac, who was well bribed for his actions), the provost of the merchants (Lhuillier), and several of the city's aldermen who arrived to open the gates and to turn over the keys to the city of Paris to their rightful king. They kissed his hand and presented him with a scarf embroidered with the royal symbol of the fleur-de-lis perhaps similar to the one he wears in Nicolay's map. Henri travelled unarmed down rue St. Honoré to calls of "Vive le Roi!" and made his way to Notre Dame to celebrate a mass and hear the Te Deum. Chroniclers reported that as the king knelt at the altar those assembled saw a small child dressed in "glistening attire" who knelt on a velvet cushion beside the king and slipped his small hand into the king's. As the Te Deum was about to begin, the apparition disappeared.[51] Those present took this site to be the king's guardian angel, and "a Divine manifestation of God's acceptance of the royal abjuration."[52] Paris was indeed worth a mass!

Henri's entry into Paris, the events surrounding it, and the message transmitted are celebrated on the map commissioned by the city.[53] The "ideological dialogue" between the humbled city and its victorious monarch is evident not only in the decorative elements but also in the geographical content.[54] Nicolay embedded this cartographical work in the political, social, religious, and military actions of the late sixteenth century and produced it as a celebratory Parisian text linking monarchy and its first city in the early seventeenth century.[55] First, in imagery, Paris is rendered a Bourbon city. The gravity of the event as it unfolded outside the city's gate along the Seine is amplified by showing France herself (a young maid in a fleur de lis dress) handing the city of Paris (symbolized by the ship) to Henri wearing the fleur de lis scarf and upon a horse who is likewise adorned with the fleur-de-lis (see figure 1.3). As Henri wearing a laurel wreath with his full armor (no helmet) reached out to *take* the city, the young woman placed her other hand on her heart in what was reportedly a Catholic gesture.[56] As Henri's horse rears up, one can see the enemies of the king and their arms lying in defeat as one Leaguer begged the king for mercy. Above the picture was a Latin phrase praising Henri's clemency, "There was no greater in piety nor in weapons of war."[57] Beneath the framed scene was a smaller frame from which hung the trophies of war and written in the frame was a quatrain in French also commending the king and his reign in Paris, "During the reign of this great king, very clement, very valiant, very just, Paris is as Rome was under Augustus, the terror of the world."[58]

The image and phrases celebrate the victory of Henri IV and the submission of Paris to this "most Christian" king. In the lower left corner, the royal privilege and the scale of measures using the French toises participated in this royal relationship.[59] Visually the location of the arms of the city, the ship with the fleur-de-lis, in the lower right corner positions it symbolically subordinate to the royal arms, France's fleur-de-lis and Navarre's golden chain, in the upper left corner where the map viewer would find the title and dedication to the king. The arms are surmounted by the royal crown, crossed olive branches and the initial H and encircling this compilation of power (and reconciliation) are the collars of the Order of St. Michael and the Order of the Holy Spirit, the Catholic order of chivalry in which Henri was the Grand Master. Bouguereau had ended the preface of his atlas with the phrase, "One God, one King, one Faith, and one Law," and pamphlets of the period echoed this sentiment and promoted the reign of this most Christian king as bringing peace and unity.[60] There was praise for his urban projects and an adoption of classical imperial imagery: just as Augustus brought peace, repose, and a fierce renown to the Rome he would rebuild, so did Henri IV intend that this well-ordered city be seen as the center of a Bourbon kingdom under his stewardship.

Second, in praise of this stewardship and ideal order as well as the royal-municipal link, the geographical display participated in the authoritative discourse. The King's Pavilion at the Place Royale—visually linked with the city's seal in the lower left corner—is grossly exaggerated to tower six stories above any of the surrounding buildings.[61] And despite the surveying of the city, Nicolay depicted the streets as wide boulevards rather than the more realistic rabbit warrens they resembled, and the city walls (especially of the Right Bank) appear secure and impassable.[62] In fact, within the walls Nicolay depicted soldiers in formation near Place St. Jacques, activity at the Horse Market near rue St. Honoré, oarsmen and boats traveling the River Seine, and Parisians going about their daily lives. Outside the walls quotidian life ensued as well, as farmers tended their fields, soldiers guarded the gate at St. Antoine, villagers and travelers rode or walked toward (and away) from the city, and the gallows at Montfaucon were occupied. Moreover, these faubourgs or suburbs were depicted as tranquil and orderly; the dense thicket of well-entrenched independent seigneurial estates were severely reduced to open and formless sprawl.[63] Henri transformed Paris into a Bourbon city and used art and architecture to shape social memories and a political consciousness that laid the foundation for what Louis XIV would do fifty years later.

Figure 1.3. Detail from Vassalieu's *Portrait de la Ville . . . de Paris* (1609).
***Source:* Courtesy of the Bibliothèque Nationale, Paris**

This would not be the last time that Paris was a hostile city for a Bourbon monarch, but it would open its doors once again for a king and even tear down its walls laying itself bare to the power of the Bourbons. The city was celebrated, but the power implicated in imagery and geometric framework was that of the crown.

By the second half of the seventeenth century, arguably, there was no better multiple media program by a European ruler than that of the Sun King, but at least two decades before he became a participatory architect of that image, he had had to rely on others. When Louis XIV became heir to the throne in 1643, he was four years old, and his mother, Anne of Austria, had to move quickly to urge the parlement of Paris both to eliminate the Regency Council created by her husband's will and to have herself named Regent. Katherine Crawford has suggested that Anne purposefully diminished her role as regent while amplifying that of her son as king.[64] In an anonymous engraving from the period and in contrast to the written descriptions of the event, the young Louis appears alone under the canopied throne facing the magistrates with Anne some distance away outside the canopy and seated next to her brother-in-law, Gaston d'Orleans (see below). "The change emphasizes the King's sole centrality. His isolation under the canopy and placement above his mother illustrate his superior status and deny any inferiority of youth."[65]

France was still in the midst of the Thirty Years' War, and the financial stress it created led to unrest throughout France and in Paris. The Regency government viewed as that of Anne of Austria and Cardinal Mazarin (the first minister), both foreigners, became easy and perhaps deserved targets of discontent over economic and political concerns. The civil unrest that followed was called the Parlementary Fronde. In early 1649, Anne fled the city fearing that there was a plot to kidnap (or rescue depending on one's point of view) the young king. The Prince of Condé led royal troops to lay siege to the city of Paris whose parlementarian Frondeurs agreed to sign a peace treaty (at the Chateau Rueil where the royal family had fled), and soon after the royals returned to the pacified city. The magistrates having acquired mostly financial concessions agreed to return to their judicial function. It is likely that the distasteful specter of the beheaded Charles I of England no doubt tempered the actions of all involved as Louis' aunt, Henrietta-Marie, Charles' widow fled to France, and no one wanted their actions to be equated with the English regicidal affair.

Nonetheless, almost as soon as the ink was dry on the Peace of Rueil, the Princely Fronde began under of all people the leadership of the Great Condé, who recently had besieged the city of Paris in the name of the Regent government. Joining Condé in leading the Princely Fronde, was Gaston d'Orléans, Louis' uncle, who with other royal princes thought the Regent government was not giving them their proper due. The arrest of Condé in January 1650 resulted in even greater hatred for the Regent's government and a resurgence of civil unrest throughout the kingdom and Paris. Mazarin went into exile in early 1650, and Anne embarked on a remarkable and dangerous political double-game maneuvering among the princes all while awaiting the day that Louis could declare his majority.

In a *lit de justice* on 7 September 1651 with great pomp and circumstance and with many Frondeurs in attendance (not Condé, even though he had been freed earlier in the year), Anne of Austria remitted the government to her son. Both would shortly leave Paris to continue their march through the kingdom to show the people their king, a stratagem that had the welcomed effect of deflating any future actions to rescue the king from the clutches of the former Regent. In Paris, while there were still many Frondeurs that rallied to this cause, by spring 1652, many in the parlement were not pleased to have the Great Condé in the city and urged the princes to make accommodations for the, after all, rightful monarch.

In March 1653, the aldermen of the city of Paris commissioned a work by Gilles Guérin. The sculpture, *Louis XIV Terrassant La Fronde*, portrayed a young Louis in Roman armor, draped in a cloak decorated with fleur-de-lis, crowned with a laurel wreath of victory, holding a baton of command, and standing with his right foot on the head of a prostrate soldier, who represented the defeated frondeurs.[66] To all who passed this sculpture in the courtyard of the *Hôtel de Ville* in Paris, even without the inscription, the message was clear: order would be maintained, and royal power would defeat insurrectionists. Paris would need to show itself worthy of its sovereign.

Jacques Gomboust's map of Paris participated in this rhetoric of dominion and reconciliation. Gomboust, a royal engineer, had begun the city project under Louis XIII and when he presented the plan to the (very) young Louis XIV, the regent government awarded him a privilege.[67] In the dedication to the king, he praised Louis' father for recognizing the necessity of such a map, but humbly presented this "very faithful, very exact, and very precise" map of "the first city of the most flourishing kingdom of all the earth" as worthy of its King "recognized to be the First and most powerful King of all the world."[68] What timing!

This map dedicated to the Bourbon monarch on the threshold of his city was bounded by a decorative garland of oak leaves, a well-chosen symbol of strength and endurance. Adorning the four corners of this map were the coats of arms of the city of Paris; the governor of Paris (maréchal de l'Hôpital); the provost of Paris (the marquis de Saint-Brisson); and the provost of the merchants of Paris or mayor by its modern title (Antoine Le Fèvre). These three men owed their municipal posts to the monarchy, grew very reluctant in the last year (or more) of the Princely Fronde to support either Condé or d'Orléans, and more than once L'Hôpital and Le Fèvre had to face or flee a Parisian mob urged on by Condé or his supporters. The map then announced from its outer edges royal dominion—like the map of Henri's Paris, it was a political manifesto for this young Bourbon king.

The expense for this map was assumed by the city of Paris, a repentant city, yet celebrated as the first in a flourishing kingdom. Gomboust boasted that the map was produced by adhering to the "rules of geometry" and was "all that is perfect in what art and the use of Mathematical Instruments can produce."[69] He presented a well-ordered map for an ordered government in which the flow of power would move in the same fashion as depicted on the map. "The dynamic maze of the streets of Paris was transformed into a neat and timeless network defined by mathematical precision."[70] In the superior right corner was the prospect of the royal palace, the Louvre, whose expanse along the River Seine from the Tuilleries to St. Germain l'Auxerrois dominated the right bank and dwarfed all other architectural compounds. In the upper left corner was the view of the city from Montmartre.

In the frame of representation of the map itself, Gomboust located two scenes with figures that participated in the "myth" of the mimetic representation of the city.[71] Just to the right of Louis' monument and bringing the viewer back into the plane of representation of the map, a horseman and his servant running behind travel towards the city but take in the view that sprawls out in front of them, a view that might resemble that from Montmartre depicted in the upper left corner. Across the River Seine beside the decorative monument (see below) in the lower right corner again the viewer is taken back into the map by two women and a servant strolling in the plain of Grenelle who similarly could see the expanse of the city before them. There are, then, three planes of representation here. Gomboust is pleased to announce the first to the map viewer who views the city from outside in an ichnographic view as though from directly above, that is, a plan of Paris. He is proud that this work was governed by the rules of geometry and suggests that unlike those that came before (no doubt Nicolay's map), his is correct. The second plane of representation is the plane in which the figures alongside each of the decorative monuments are displayed. They serve a deictic function as if to say to the map viewer: "behold the city of Paris." These figures do not see the city of the map viewer, that is, the ichnographic view, but a view more akin to that depicted in the upper left corner, the view of Paris from Montmartre, or that in the upper right corner, the span of the Louvre from the left bank. This relationship gives credence to the sense that the map is a snapshot, a mimetic representation, a "true" representation of the city and that is what Gomboust wants the reader to believe. Finally, the third plane of representation is the traditional bird's eye view which though banished from the city plan is evident in the depiction of the various royal chateaus and princely homes surrounding the city of Paris. One can view the city while viewing the chateaus that surround the city and, in effect, Louis could view Paris, his flourishing city devoted to him, while around the city the chateaus were displayed as though in humbled procession or in audience with the king (see figure 1.4).

Fourteen of these views of royal residences adorn the two substantial monuments (or pedestals) that anchor the work in a sphere of patronage. The sturdy pedestal in the lower left corner (just below the dedicatory address to

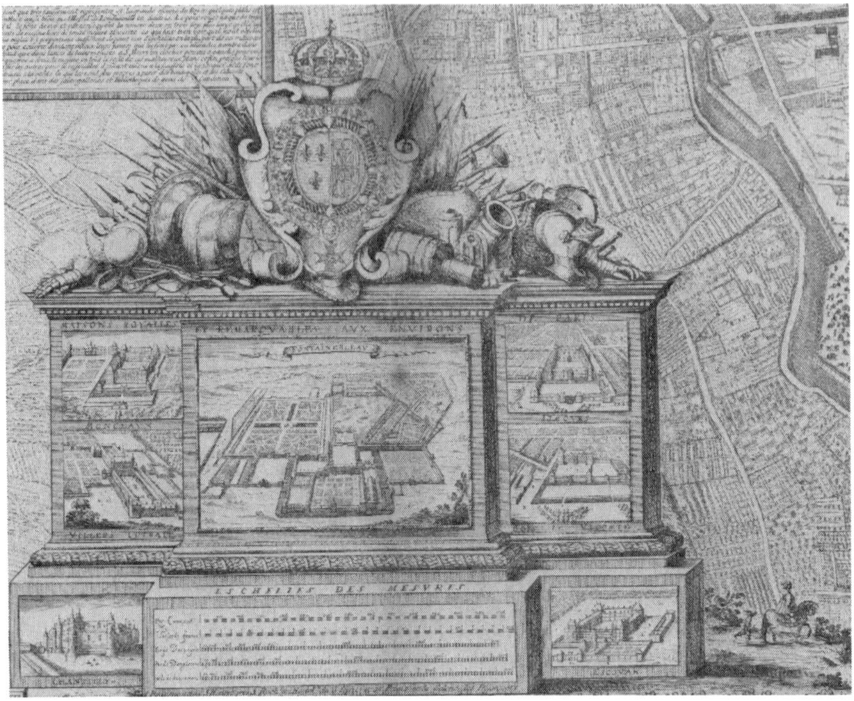

Figure 1.4. Detail from Gomboust's *Lutetia, Paris* (1653). *Source:* Courtesy of the Bibliothèque Nationale, Paris

the king) includes the royal coat of arms flanked by the military arms and trophies of Louis XIV. In the right corner a slightly smaller pedestal was topped by the coat of arms of Pierre Séguier, Louis' chancellor and keeper of the seals, whom Gomboust praised for his support of the project in the address to the reader located just above the monument. These large foundational structures reflected the efforts and intentions of this young monarch to build a strong foundation for France after the upheaval of the Fronde(s).

Louis Marin has suggested that Gomboust's pride in his geometrical presentation was born of his confidence in the mimetic nature of geography.[72] Here is Paris, that is, by the rules of mathematics, this really is the city of Paris. The reality displayed, however, is just the constructed reality of the moment. The work is a lovely ichnographic view, but by understanding the historical and social context of the map, the view is enriched whether it is the supposed mathematical clarity or the cultural iconography. Many of the fourteen bird's eye views displayed, such as Fontainebleau, St. Germain en Laye, Bois-le-Vicomte, Limours, Rueil (where peace was signed with the Parlementarian Frondeurs), Chantilly, and Vincennes to name a few, attest to courtly intrigues with mistresses and financial mismanagement. Chantilly, the chateau in the family of the Great Condé no doubt offered a cautionary tale to its sometime resident, frondeur, and current mercenary to Spain. (In 1632, Louis XIII had its previous owner beheaded, the marshal of France from the Montmorency family, for treasonous activity.) Beneath these monuments along the bottom of the map were another ten noble houses some belonging to parlemenatarians, financiers, etc. some of whom had clear frondeur connections such as the duke of Vendome, the "natural" son of Henri IV at chateau d'Anet and duke of Chevreuse at chateau de Dampierre. In every corner, Paris bowed to their Bourbon king. These chateaus carried the imprint of royal power and radiated it out into the surrounding area, but so close to the seditious actions of the high nobles they (as if in procession) now stood as obedient subjects waiting to be attentive and dutiful adornments to the young man who would become the Sun King.

CELEBRATING THE COMMERCIAL CITY OF LONDON

A very different city map appeared in London in the late sixteenth and early seventeenth century. A near contemporary of Henri's royal engineer, Nicolay, John Norden (c. 1547–1625) an Oxford-educated surveyor, cartographer, chorographer, and writer of devotional works, embarked on an endeavor, *Speculum Britanniae* or *Looking Glass of Britain*, to expand upon Saxton's county map project.[73] He would not only survey the counties to include roads but also key locations and characteristics noted by symbols on his maps and provide the chorographical descriptions missing in Saxton's atlas. Although he was less successful in acquiring more lucrative patronage from the Tudor government and his *Speculum* project languished and remained incomplete, he did achieve a royal appointment as surveyor of the Crown woods in 1600.[74]

When Norden's plan of London, the first of its kind, appeared in 1593 it was one of three maps in the first part of his *Speculum* project covering the county of Middlesex. Elizabeth's chief minister, William Cecil, Lord Burghley, had seen Norden's earlier manuscript for Northamptonshire as had Christopher Hatton, both members of Elizabeth's Privy Council, and Norden was soon given permission for publication of his work. This support, to Norden's displeasure, was not monetary, but he proceeded with his project. Closer to home, he initiated his description of Middlesex and again presented it to Burghley. It was the only portion of the *Speculum* to appear in print and its frontispiece not only displayed the Tudor arms and motto, but linked the two figures, surveyor and patron. Frank Kitchen notes that the figure of Burghley appeared with his hand in his pocket as though to offer the support Norden so consistently sought.[75] Within this "historicall & chorographicall" description of Middlesex was Norden's plan of London which spanned from East Southwark to Gray's Inn and north from Islington to Southwark across the River Thames (see figure 1.5). Along the bottom of the map was a list of streets and buildings so noted to correspond with numbers and letters on the map for locational purposes. This addition was unique at the time and set Norden's work apart as truly useful at the ground level. And yet coupled with this utility was an ideological message that suggests a parity of the City and the Crown. The livery companies of the City asserted their privileged independence in the culture of the city and viewed themselves as complementing royal power for the commonweal.[76] Along the top center was a scroll with the title of the map, London, and to the left was the royal coat of arms and to the right the arms of the City of London. These emblems sat along the same representational plane and linked the royal authority and the municipal authority neither overwhelming the other, and along either side were the arms of the twelve principal livery companies of the City, such as the Drapers, the Fishmongers, the Merchant Adventurers, and the Goldsmiths. The City speaks of its commercial life as the vessels sail or row along the River Thames and the viewer reads from the list along the bottom: Y The Exchange; Z Cornhill; 5 Leaden Hall; 18 Olde Baylye, and so on.

The 1593 version of this map slightly smaller than its six inch by eight inch original size without the arms of the twelve London companies and without the royal and city arms appeared as an inset on Norden's large *Civitas Londini* (1600) which reaffirmed the municipal authority. The small plan of London was juxtaposed to the large sweeping view of the city which included the names of prominent London sites along the skyline. Opposite the plan of London and in the bottom left corner was a plan of Westminster, a map unfolding upon the land, the seat of royal presence and directly above in the upper left corner were the crowned Tudor arms with motto and supported by the lion and dragon. Above the plan of the City were the arms of the City crowned with a laurel wreath and a pair of banners, one of the city arms and the other of the arms of the current mayor. The scene below the map depicted a view of one part of the Lord Mayor's pageant as he made his way through the City.[77] The coats of arms lie along the same representational frame and now linked as they would be in seventeenth-century maps are the cities of London and Westminster. The textual insert linked royal and civic as well: "This description of the moste famous citty, London, was performed in the yeare of Christe, 1600, and in the yeare of the moste wished and happy raigne of the right renowned Quene Elisabeth, the fortye and two. Sir Nicholas Moseley Knight being Lorde Maior and Roger Clarke and Humphrey Wylde sherifes of the same."[78] London may have been a royal city like Paris, and its occupants more than once sorely tested its monarch, but, it was never in the shadow of Westminster. Unlike Paris, London's municipal and com-

Figure 1.5. Norden's *London* (1593). Facsimile from the London Topographical Society, 1899. *Source:* Courtesy of the Yale University Map Library

mercial power was always on display. When city and monarch linked, it was with a mutual recognition of power.

For many the happy reign of Elizabeth had lasted too long, but there was a mixture of hopefulness and trepidation for the accession of James VI of Scotland to the throne of England as James I. Almost immediately Norden sought the patronage of the new king and enjoyed the attention if not the patronage of James I and Lord Burghley's son, Earl of Salisbury. On account of his surveying talents, he acquired several private commissions for surveying work.[79] In 1607, he dedicated a royally commissioned work, *A Description of the Honor of Windesor* [sic], to James. The manuscript on vellum included descriptions and a view of the castle, a map of the forest, and fifteen maps and plans of the parks within the area.[80] Similarly, Norden presented ten manuscript maps covering the county of Cornwall in his bid to acquire patronage for his Saxtonesque project, *Speculum Britanniae*.[81] Patronage on the scale sought by Norden simply was not forthcoming, an apparent disconnect with the great fanfare with which James was greeted at the royal entry in March 1604.

The ten-month delay due to plague and the king's rather "sullen disregard" for such events that took him from the hunt had not dampened the grand spectacle of the 1604 procession. In imagery and verse the power of the monarchy was praised and reasserted the Elizabethan iconography of a golden age. At the sixth triumphal arch set up at Fleet Street, Zeal spoke of "the populous globe of this our English isle" which had appeared to move backward since the death of Elizabeth, but now with the "peaceful presence" of this new sovereign, "our globe is drawn in a right line again."[82] The implication was that James would restore the golden age; he was the new Augustus and London, the new Rome. Emphasizing the relationship between the city and crown, Zeal concluded his speech, "Of this glad city, whither no Prince ever came / More loved, more longed for, lowly I entreat / You'd be to her as gracious as you're great."[83] Here, the rhetoric was for a hopeful benevolence (toward the city) proportional to a hopeful magnificence (of the king).

At the first triumphal arch (financed by the City of London) to greet the royal entourage, the city presented a tentative and uncertain relationship. The arch was crowned with an eastern prospect of the city similar to the one on Norden's *Civitas Londini*. In the stage below at the center sat the figure of Monarchy who held a scepter in one hand and a globe in the other, and at her feet was the figure of Wisdom by whose countenance and conduct great kingdoms arose. In separate panels to the left and right of these figures were the coats of arms of James I and the City. While the adapted panorama sat hierarchically above the figure of Monarchy, the armorial bearings appeared on an equal representational frame. Perhaps more descriptive of its status was the inscription located at the base of the stage and spanning the entire arch, *par domus haec coelo, sed minor est domino*. (This house [i.e., London] is on a par with the heavens, but less than its master.)[84] In other words, the city may rise towards the heavens, but it did not surpass its sovereign. In closing his speech to the spectators at this first stop on the processional route, the Spirit (Genius) of the City reminded that the city could still provide for the crown and hoped for the continued advocacy and support from the king.

It appears many were to be disappointed, as within the last decade of James' reign, London (and England) faced economic depression, the events of the Thirty Years' War (1618–1648) threatened to draw England into European confrontations, and Parliament (more accurately, the Commons) presumed—according to James—to meddle in affairs (e.g., the Spanish match and the restoration of the Palatinate) beyond their privileges. Parliament granted several subsidies for the purposes of preparing for war, but James had to concede that the distribution of the funds and any peace negotiations would be through consultation with the members. Of the two new Augusti—Henri IV and James I—and the two new Romes—Paris and London—in imagery (and in power), James would not dominate the scene of London. As mentioned above the monarchy had already been *displaced* from the center of the map and the crown was in no position to provide patronage for cartographic projects on the scale of Saxton's atlas. In fact, commercially Saxton's maps continued to be copied or reissued with few if any revisions for several more decades.[85]

This practice is even evident by Norden who re-issued his 1593 map thirty years later. In 1623, he added to the re-issue a strip along the bottom that included a second title, *A Guide for Cuntrey Men. In the famous Cittey of London by the helpe of wich they shall be able to know how farr it is to any street . . .*, and increased the number of streets and places to 95. London had grown, and anyone arriving to London would have that many more locational aids as he made his way through the great commercial city of London. It was quite a popular map in its leaflet form and was continually published into the second half of the seventeenth century. The royal arms were still present along the same plane as those of the City, but they remained the Tudor emblem rather than those of James I, the Stuart king for, by then, two decades. Was this simply a practice of the trade? Or was there really no compelling reason to celebrate or acknowledge parity of the City with the Crown? Perhaps the answer lies in both elements.

Thirty years later after Norden's death and the sale of many of his copperplates to Peter Stent, a London print and map-seller, the map was re-issued again (by Stent) under the Commonwealth with little change to the 1623 version except the date, the inclusion of Stent's address, and the elimination of the name of Pieter Van den Keere as the engraver.[86] One might argue that Cromwell and supporters of the Commonwealth did look fondly upon the days of "Good Queen Bess," but it would be a stretch to suggest that Stent did not update the coat of arms for that purpose. It was a useful map that would still sell and the very reason for Stent to purchase the copperplate in the first place. If the information on the map was still useful why waste any of the profits on revising the copperplate.

MANIFESTING THE PROSPERITY OF ENGLAND

Decades later, however, after political turmoil and an act of regicide, the restoration of Charles II to the throne brought such optimism for the peace and prosperity that this rightful monarch could bring that there was briefly great fanfare. All were well aware of the nearly two decades of upheaval and all wished to move forward and hoped to see in this king a return to moderation and the banishment of extremes. A participant in and director of the restoration festivities and fanfare was John Ogilby (1600–1676), no stranger to second lives and new opportunities. This former dance master turned publisher became royal cosmographer and produced the celebratory procession that took place in the streets of London and published it in 1661.

While the reign of this Stuart king may not have had the ability to advance or even attend to the enduring pageantry of the Elizabethan era, the rhetoric of the early years of this period was optimistic and did suggest that he was a new king come to establish order to a fractured kingdom.[87] Some of this very rhetoric could be found in Ogilby's work for the coronation of Charles II entitled, *The Relation of His Majesties Entertainment Passing Through the City of London to his Coronation* (1661).[88] As a former dance master and translator of several classical texts, Ogilby was well suited to the display and its description steeped in classical motifs. In the description of the first triumphal arch along the processional route, Ogilby described the representation of Monarchy situated on the south pedestal before the arch as a figure wearing the majestic purple robe which was adorned with diadems and scepters. Worn by Monarchy, the loose mantle with a depiction of the map of Britain upon it, enveloped her figure. She was supported by Loyalty who carried three crowns and three scepters corresponding to her sovereignty over the three cities: the depictions of London upon her head, Edinburgh in her right hand and Dublin in her left hand. This geopolitical rendering with monarchy carrying or supporting Britain and holding sovereignty over the three royal cities recognized the dominion of the Stuart king over the Isles but does not quite return to the Elizabethan imagery of the personal monarch in a position of dominance of the land.[89] In these heady days of the early Restoration, Charles may have been the savior of the people, but he was also in service to them, that is, "our laws, liberties, and religion." Charles was to adhere to the tradition of "king-in-parliament" and limited government and would certainly be wise to avoid the absolutist trappings of the Bourbon king, Louis XIV. But England was still early in the honeymoon stage of this reign. In his second expanded edition on Charles' coronation celebrations, Ogilby wrote of Charles' triumphant return: the "glorious Restauration of our Sovereign to His Throne, and of us His Subjects to our Laws, Liberties, and Religion, after a dismal Night of Usurpation, and Oppression."[90]

Before moving on to the second triumphal arch in the coronation ceremony, this figure of Monarchy responded to a speech by Rebellion—depicted on a north pedestal before the arch as a woman on a Hydra with a crown of fire and snakes in her hair. Monarchy exclaimed: "To Hell, foul Fiend, shrink from this glorious Light, And hid thy Head in everlasting Night." As she turned to welcome Charles, Monarchy exclaimed: "Enter in Safety, Royal Sir, this Arch, And through your joyful Streets in Triumph march; Enter our Sun, our Comfort, and our Life."[91] At the center of the arch an image of Charles on horseback in "calm motion" witnessed Usurpation fleeing before him into the jaws of hell.

Sadly it was shortly thereafter that London itself appeared to be nearly consumed by the jaws of hell as fire tore through much of the city. Ogilby still relatively new to his career in publishing had found success and new patronage with his translations and publications of the works of Virgil and Homer. In the Great Fire of 1666, this taste of success soon went up in smoke as his new publishing business was destroyed with so many other businesses and homes. One is reminded of James I proclamation in 1610 regarding "restraint of building in and about London," but efforts to control the urban sprawl were to no avail.[92] The Lord Major and Aldermen of the City chose Ogilby along with his (Ogilby's) step grandson, William Morgan (d. 1690), as sworn viewers with the power to arbitrate the expected property and boundary disputes that arose after the Great Fire. Ogilby found himself making connections not only to key members in the rebuilding effort such as Robert Hooke, a member of the Royal Society, but also skilled surveyors such as William Leybourne who had with five others been chosen to survey the City.[93] These connections would prove crucial for the success of his surveying (see chapter 2) and his subsequent road atlas.

Ogilby's *Britannia* (1675), one of the "landmarks in British map-making" and the "first English road atlas," was to be a testament to geographical knowledge in service to the state and reflected the economic and social optimism of the Restoration period.[94] In late 1669, Ogilby had earned a fifteen-year privilege for a rather ambitious geographic project, a five-volume work of the world, an atlas for each of the four parts and a grand atlas devoted to the "Business of Great Britain."[95] While the grand project was never fully realized, with his *Britannia*, Ogilby offered, by subscription, an atlas of all the principal roads for England and Wales as a depiction of Charles' flourishing kingdom and was promised 500 pounds each by the king and the queen consort.[96] Moreover, Ogilby, royal cosmographer since 1671, wrote in the dedicatory page: "I might be bold to tell your majesty, that we, who have hitherto seem'd to truckle under the fame of France and Belgium, for performances of this nature, may presume, by the propinquity of this our design to perfection, that perchance no attempt at first came nearer, to have given such measures to the virtuosi of the world, as foreign princes and states shall be glad to imitate."[97]

Wenceslaus Hollar, a famous seventeenth-century engraver who also worked on Ogilby's *Coronation* (1661), was called upon to produce the visual introduction for this road atlas (see figure 1.6). The scene was the gate of London topped with the royal flag, the River Thames in the distance, and the roadway sprawling out to the countryside as travelers, horsemen, carriages, and wagons made their way to and from the city. Just outside the gate two horsemen studied a strip map before setting off and three other men gathered around a table scattered with measuring and surveying devices and contemplated the globe while in the distance a group of men used the perambulator reflecting the surveying method (road traverse) adopted by Ogilby to create the road maps. Above the scene little cherubs hovered holding a banner with the atlas title and three strips maps much like the ones found in the atlas. The scene was one of tranquil activity and social order as travelers went about their business well-aided by road maps. In his dedication, Ogilby offered "I have attempted to improve our commerce and correspondency at home, by registering and illustrating your majesty's highways, ... I present your majesty with an important novelty, the scale of peace and war, whereby . . . a true prospect of this your flourishing kingdom may be taken, pregnant hints of security and interest."[98]

Not only was this work a testament to the thriving and stable kingdom, but also the fruition of the coronation rhetoric whose "Sun," "glorious light," and "comfort" had made possible the prosperity, tranquility, and ingenuity depicted in the frontispiece. This rhetoric complemented the recent truce (short-lived) between Charles and Parliament, but belied the tense realities which would bring about the Glorious Revolution in the following decade. Domestic issues aside, nonetheless, this atlas was to be a British showpiece. No foreign rulers had produced such a work though many would, Ogilby claimed, "be glad to imitate." Royal support was linked with national achievement and divine approbation. "May the same influences tend to be the culminating of all the other arts and sciences, and the King of Kings crown your sacred head with peace and plenty, raising up the splendor of Great Britain's Monarchy, and rendering it the grand exemplar of industry and ingenuity."[99]

In the year of Ogilby's death (1676), Morgan published an ichnographic plan (in twenty sheets) of the city of London from the Tower to Lincoln's field.[100] This "first class" example was a magnificent civic production.[101] The map carried a dedication to the Corporation of the City (Lord Major, Aldermen, and sheriffs) that had given Ogilby the permission to carry out the survey and prominently displayed the arms of the City and the emblem of the then current mayor, Joseph Sheldon (see figure 1.7). On the map then the imagery was civic authority, and the royal presence was only acknowledged in the twenty-first sheet carrying the dedication to Charles II. In a recent article, David I. Bower suggests that the map was intended to be quite a bit more grand with the arms of the city and the arms of the king prominent along the top center border and to the left and right of them the arms of the twelve major livery companies, vignettes of prominent London buildings along the sides of the map, a border along the bottom of the map with the arms of the sixty remaining companies, and a panorama of the city below that border.[102] After Ogilby's death in September 1676, however, Morgan took to heart Ogilby's concern over plagiarists and rushed to print the map with much less ornamentation, so much so that the Crown appears to have slipped off the map. Bower also acknowledges the earlier comments by Ralph Hyde in offering that the lack of decoration may also have been due to the aldermen's reluctance to give generously for the production of the map.[103] With the homes of prominent merchants noted on the map and Sheldon's arms

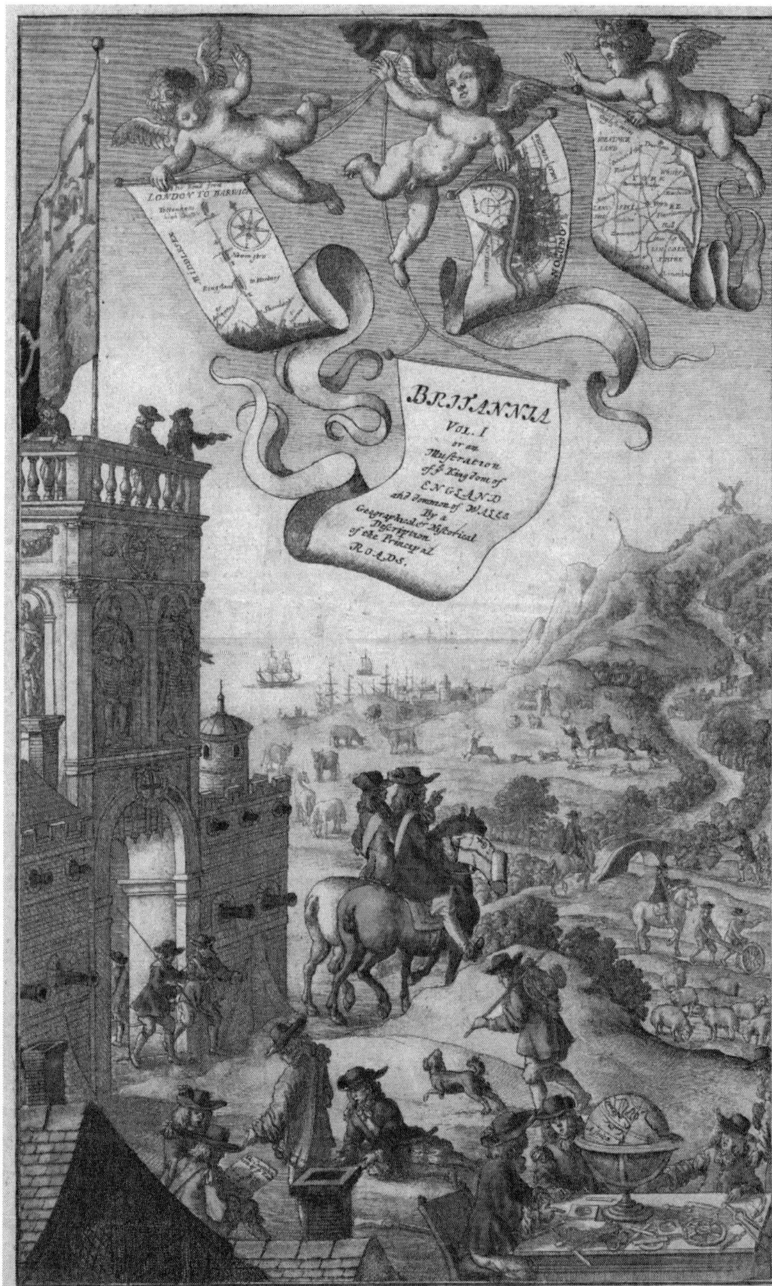

Figure 1.6. Frontispiece from Ogilby's *Britannia* (1675). *Source:* Courtesy of the Beinecke Rare Books and Manuscript Library, Yale University

(rather than the king's) prominently displayed "in order to please the paymasters," this was a "City map."[104]

Six years later, however, Morgan published a new map of London which was, as Peter Barber noted, "blatantly royalist."[105] The map carried an equestrian portrait of Charles I in the upper left corner and a similar depiction for Charles II in the upper right corner with the exception that Charles II's horse is trampling a figure, perhaps the Earl of Shaftesbury or a representative of the Whigs. In 1681 Charles was victorious over the Whigs regarding the Exclusion Bill which sought to prevent James II (a Roman Catholic) from the throne. He had dissolved Parliament, chastened the City of London reminding it that it held its franchise and liberty by the grace of the King, and allowed more Tories to take leadership roles in government. Royalist celebrated.

Morgan celebrated as well, and in the dedication to the king wrote: "May the great God long preserve your Majestie and suddenly convert or confound all your enemies." He praised Charles' action: "Let the world see how able your Majestie is to befriend your Allies and punish your enemies[,] succour

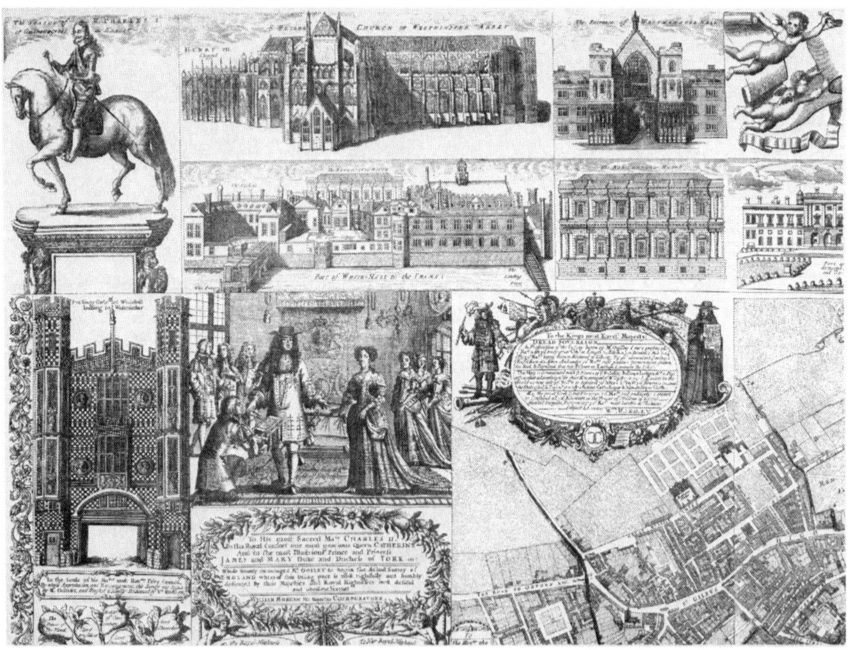

Figure 1.7. Detail from Ogilby's London (1677/1682). Facsimile from the London Topographical Society, 1904. *Source:* Courtesy of the Yale University Map Library

the distressed and defend ye truly antient Catholique and Apostolique faith."[106] The rhetoric being promulgated in speech and text was that Charles had averted civil war and his governance was praised. Similar to the praises in Ogilby's *Britannia*, Charles' governance "banished extremes" and brought prosperity. For Morgan and Ogilby there was great appreciation for the royal patronage as joining the imagery of the equestrian statues, the views of Whitehall, St. Paul's, St. Peter's, and the Royal Exchange was the image of Ogilby kneeling before the King and Queen presenting his subscription list for the survey. Morgan had completed the survey of London and depicted a larger area for the city from St. James' Park to the pasture fields beyond the Tower of London. Joining these images were lists of the names of individual subscribers from the Lords to the Commons and such prominent corporations as the City livery companies and the universities. Here was a city that had grown and prospered, and given recent tensions emerged unscathed and thanks to its king, a city like no other. The work of Ogilby and now Morgan carried forward the praises of the early days of the Restoration, but with Charles death in 1685 and the Glorious Revolution of 1688, the Morgan map of London would reappear in 1690 stripped of all royal imagery and highlight, once again, the civic elements and commercial presence of the city.[107]

MANIFESTATIONS OF MAGNIFICENCE IN FRANCE

In the early period of Louis' personal reign (1661–1715), the government or state was centralized around a powerful image of the king thus confirming the personal state, *l'état c'est moi*. In what has been called a department of glory, Jean-Baptiste Colbert (1619–1683), director of the king's projects and finance minister, charged a group of leading and influential artists, writers, architects, and musicians to produce works to contribute to the public image of the king and Bourbon *gloire*. These works aided in the fabrication of the cult of the Sun King.[108] Mapping was a viable concern in this department of glory.[109]

As Ogilby was planning his grand multi-volume atlas project in the late 1660s and early 1670s, Alexis Hubert Jaillot, a sculptor from Franche-Comté turned map editor, had begun publishing maps and in 1670 initiated a business relationship with the sons of Nicolas Sanson (d. 1667), royal geographer to Louis XIII and Louis XIV.[110] It lasted long enough for Jaillot to publish his famous *Atlas Nouveau* (1681), a compilation of the individual maps published by him and sold separately in the preceding decade. The timing was auspicious as Louis XIV had declared war on the United Provinces beginning the Dutch War (1672–1678). Jaillot dedicated many of the individual maps to the young Louis, the Grand Dauphin, and then the entire atlas to the now twenty-year-old heir to the French throne.

The architectural frontispiece of the atlas celebrated the triumphant Bourbon arms.[111] The structure is reminiscent of Bouguereau's frontispiece praising Henri IV, the triumphant arches of Paris such as Porte St. Denis, and the celebratory imagery of the early years of the personal reign of the Sun king (see figure 1.8). Louis XIV and his son were presented in Roman dress on horseback surrounded by armor and the implements of war. The equestrian model, a classical form borrowed from Renaissance art, had already been used in Pierre Mignard's work, *Louis at Maastricht*, to celebrate Louis' victory in 1673.[112] Here in this frontispiece was the presence of the king now just over forty years old and joined by the young man that would rule after him (or so it was thought). These were no mere symbols of Bourbon rule but the personal image of the king and his son. The classical theme continued as Hercules, Mars, and Pallas Athena joined in celebration of Bourbon glory. Completing the war motif and complementing the warrior helmets that rested along the top of this architectural structure were bas-reliefs, at the base of the façade, depicting two of the three major battles of the Dutch War, Seneffe, a tragic pyrrhic victory after the success of Maastricht, and Cassel, a victory in the last year of the war.

With this work Jaillot established his reputation as a premier map editor and having sought to align his map editing business with the interests of

Figure 1.8. Detail from frontispiece of Jaillot's *Atlas Nouveau* (1684). *Source:* Courtesy of the Bibliothèque Nationale, Paris

power and influence, he succeeded in 1686 as he was awarded the title of royal geographer.[113] Although without the benefit of his business partnership with the Sansons, Jaillot soon embarked on a bold project to produce a two-volume work, one to cover the world and the other to cover France. His *Atlas Français* published in 1695 was the only product of this grand design, and although approximately half of the maps of this atlas were reduced versions of those appearing in his first atlas, he included for posterity a portrait of himself, a mark of his distinction in the field. He humbly placed this atlas of the world at the feet of his king, the ruler of the civilized world. While Jaillot's hyperbole and imagery would indicate one glorious scenario, the events and history revealed another. By the publication of the *Atlas Français*, France was nearing the end of the War of the League of Augsburg (or Nine Years' War) and the victories it savored and celebrated were difficult to find, and finances were more than strained as the silver furniture at Versailles was melted down for currency.

In the final decades of his reign (1643–1715), Louis' image was tarnished and reflected more the image of the god Phaethon, who lost control of the sun chariot, than the image of the powerful sun god, Apollo.[114] However, in the frontispiece of the *Atlas Français*, Jaillot presented a Herculean vision of Louis XIV as Atlas aided this French Hercules in *righting* upon his shoulders the globe, which carried a faint outline of Europe with France facing the viewer and the royal *fleur de lis* engraved on the globe (see figure 1.9). Victory crowned the globe with a laurel wreath as Glory trumpeted French triumphs.[115] The personifications of Asia, Africa, America, and Europe looked upon the scene with awe and perhaps terror. Strewn about were the implements of war and the atlas title is printed upon a ship's sail held by a cherub and Europe. In the background a ship carrying a fleur-de-lis banner not only signals France's naval presence against its enemies, but also its challenge of, in particular, British merchant trade. Despite the hyperbole, the image was clear; the world was a reflection of the theater of the glory of France, still. In Jaillot's dedicatory text, Louis, by the magnificence of his reign, his great conquests, and his wisdom, filled the world with admiration, terror, and veneration.[116] In image and in word, though a bit overstated, France ruled the world. Its magnificence, however, was in the hands of the personal monarch.

This same imagery and hyperbole was reflected in a large map of France (1693) engraved and sold by Jean Baptiste Nolin, engraver and map seller in Paris[117] (see figure 1.10). This map was advertised in the *Journal des Scavans* whose editors described it as a magnificent work.[118] The imagery as with Jaillot's atlas frontispiece above was so necessary as the intended short and victorious plan for the War of the League of Augsburg was not to be as its other title indicates, the Nine Years' War. Louis' inability to be decisive when the war began to languish and instead adopting a plan of waiting out

Figure 1.9. Frontispiece from Jaillot's *Atlas Français* (1695). *Source:* Courtesy of the Beinecke Rare Books and Manuscript Library, Yale University

the enemy required a public image of power and domination over the perceived enemies.[119] The assembled group depicted on the map included the King sitting in great repose to emphasize his tranquility in the midst of the most furious actions of war. He was dressed in his armor holding the baton of command and defended Piety, Justice, and his allies as he crushed beneath his foot the hideous figure of Heresy identified in the *Journal* as the inevitable cause of the troubles over which Europe was disturbed. The editors of the *Journal* certainly would not suggest any cause connected to the actions taken by Louis XIV to solidify his reunions of the 1680s.[120] Renown in the background raised the curtain on the evil designs of Louis' enemies who were overwhelmed by the figures of Piety and Valor, attributes that had been earlier praised in the actions of Henri IV, Louis' grandfather and the first Bourbon king. Accompanying Louis in his assembled council were the figures of Time, Geography, and a seated Hercules looking upon a map of England whose coastline more than once had been threatened by the French fleet. It is of little concern in this imagery that these actions against England had taken French finances to the very brink. Once again, the celebratory rhetoric was the glory of the Bourbon crown.

Even Nolin's map of Paris (1698) participated in the *gloire* of the Sun king at a time when the sun was in its descent[121] (see figure 1.11). Similar to the dedication on his 1693 map of France, Nolin associated Louis' military achievements along the boundaries of France with the approbation and gratitude of his Parisian subjects. In his extravagant dedication on the map, Nolin declared that just as the lands that enjoyed the warmth of the sun the most were the happiest and most abundant in all things, Parisians, so close to the magnificence of Louis, were the happiest in the kingdom. Furthermore, Paris, as the seat of his justice and power, reflected his glory just as his conquests around the borders of France reflected his military strength. The imagery on the map was no less laudatory as the sun symbol emblazoned with Louis' motto, *Nec Pluribus Impar*, was surrounded by the symbols of the zodiac placing the Sun king not merely in the terrestrial realm but the celestial realm. Beneath the structure on which this cosmological display was engraved, were the allegorical figures of Government, Prudence, Fortitude, and Wisdom, all of whom were, in reality, sorely missing at that time in Paris. As with Jaillot's frontispiece to the *Atlas Français*, Nolin's decorative cartouches present the historian with a purposeful or designed disconnect between the imagery on the map and the reality on the ground.

CONCLUSION

By the late seventeenth century James II had succeeded to the throne on the death of his brother, Charles II. It was not long before James' Anglican

Figure 1.10. Detail from the cartouche of Nolin's *Le Royaume de France* (1693).
***Source:* Courtesy of the Bibliothèque Nationale, Paris**

subjects grew alarmed at his decisions, and the tensions between the Crown and Parliament over his support for Catholicism and attacks on opposing bishops reached a breaking point. The king's demands on Parliament to fund his standing army and his growing disregard for the political position of Parliament, which led him to adjourn the body in November 1685 and to dissolve it in July 1687, set the stage for actions against the king. In June 1688, soon after the birth of his son, a Catholic heir and future James III, Anglican leaders sent an invitation to William of Orange to invade England. James fled the country and arrived in France on Christmas day. The transition of power or Glorious Revolution gave birth to a constitutional monarchy as a Convention met in January 1689 to declare that James II had subverted the constitution and had broken the contract between the King and his people. In February 1689, the House of Lords and House of Commons offered the crown to William and Mary. Almost immediately William declared war

Figure 1.11. Cartouche from Nolin's *Les Environs de Paris* (1698). *Source:* Courtesy of the Bibliothèque Nationale, Paris

on Louis XIV, a condition that plagued the last Stuart monarchs into the early eighteenth century and further allowed the Parliament to solidify its powerful political position within the constitutional monarchy. The Revolution brought the transition from the personal rule so ably represented by the Tudor reigns of Henry VIII and Elizabeth I with flashes during the reign of Charles II to a parliament-centered rule. Neither William (nor Mary) participated in a multiple media program of Stuart dominance which would have not only been unseemly but caused great alarm in Parliament, especially to the gentry in the Commons.

For Louis, the English form of government was at best distasteful, but in the last two decades of his very long reign (seventy-two years), his personal rule was plague with wartime debt, social crises, and few events to prompt celebrations of his glory. Peter Burke refers to this time period as the "royal sunset" in which the previous architects of the cult of image of the Sun King were all dead, and many artists, engravers, writers, etc. fled France after the Revocation of the Edict of Nantes (1685).[122] Court and state patronage to celebrate Bourbon *gloire* was out of place in the current political, economic, and military environment.

By the late seventeenth and early eighteenth century the Renaissance-esque personal rule, which had held on a bit longer in France, was giving way to the state-centered government that in England required recognition of the various elite and corporate bodies such that there was little centralization or institutional infrastructure to direct the expensive and organizationally demanding projects for the Enlightened state of the eighteenth century. In France McClellan and Regourd have identified a Colonial Machine, a "bureaucratic core and a constellation of dependent specialized institutions"[123] whose directing gaze was towards the colonies but whose many institutions had domestic responsibilities. Maps with or without decorative elements served the needs not only of England's elite and corporate bodies, but also those institutions (of the Machine) with interests within the borders of France.

In cartographic productions, there was no less art on the geographical works of the eighteenth century, but baroque sensibilities had declined, cartographers wanted to appeal to the Enlightenment's scientific discourse, and the personal state was no more. Louis' astronomers working on the Paris meridian were able to show him the inaccuracies of the maps of France. While he was surprised that his astronomers had lost him more land than any of his generals, it was evident that science was useful to the bureaucratic state he had created. His geographers and map makers who had served the personal state with valuable works and elaborate cartouches to celebrate his *gloire* would now serve this bureaucratic state with cartographic productions drawn from empirical information and reasoned analysis of sources. As science was in service to the state, a rhetoric of power is still evident on the map but is without the *gloire* and often intertwined with a rhetoric of scientificity. The chapter that follows takes up the investigation of the regional and county maps of England and France from the seventeenth to the eighteenth century.

NOTES

1. See Christian Jacob, *The Sovereign Map: Theoretical Approaches in Cartography throughout History*, trans. Tom Conley, ed. Edward H. Dahl (Chicago: University of Chicago Press, 2006); Louis Marin, *Portrait of the King*, trans. Martha M. Houle (Minneapolis: University of Minnesota Press, 1988); and J. B. Harley, "Deconstructing the Map," in *The New Nature of Maps: Essays in the History of Cartography*, ed. Paul Laxton (Baltimore and London: Johns Hopkins University Press, 2001), 149–96.
2. Mary Sponberg Pedley, *The Commerce of Cartography: Making and Marketing Maps in Eighteenth-Century France and England* (Chicago: University of Chicago Press, 2005), 63.
3. See Roy Strong, *The Tudor and Stuart Monarchy: Pageantry, Painting, Iconography*, vol. II (Rochester, NY: Boydell Press, 1995), 14–15; and René Graziani, "The 'Rainbow Portrait' of Queen Elizabeth I and its Religious Symbolism," *Journal of the Warburg and Courtauld Institutes* 35 (1972): 247–59.
4. Stevan Dedijer, "The Rainbow Scheme: British Secret Service and Pax Britannica," in *Clio Goes Spying: Eight Essays on History of Intelligence*, eds. Wilhelm Agnell and Bo Huldt (Lund: University of Lund, Sweden, 1983).
5. Strong, *Tudor and Stuart Monarchy*, 14. Also see Daniel Fischlin, "Political Allegory, Absolutist Ideology, and the 'Rainbow Portrait' of Queen Elizabeth I," *Renaissance Quarterly* 50, no. 1, (1997): 175–206.
6. Strong, *Tudor and Stuart Monarchy*, 13–14.

7. Strong, *Tudor and Stuart Monarchy*, 13.

8. See Peter Burke, *The Fabrication of Louis XIV* (New Haven, CT: Yale University, 1992); and Peter Barber, "England II: Monarchs, Ministers, and Maps, 1550–1625," in *Monarchs, Ministers, and Maps: The Emergence of Cartography as a Tool of Government in Early Modern Europe*, ed. David Buisseret (Chicago: University of Chicago Press, 1992), 77–84.

9. T. C. W. Blanning, *The Culture of Power and the Power of Culture: Old Regime Europe 1660–1789* (Oxford: Oxford University Press, 2002), 7.

10. Kevin Sharpe, "Representations and Negotiations: Texts, Images, and Authority in Early Modern England," *The Historical Journal* 42, no. 3 (1999): 870.

11. Strong, *Tudor and Stuart Monarchy*, 10.

12. Elizabeth I, "Queen Elizabeth I of England, Selected Writings and Speeches," in *Modern History Sourcebook*, ed. Paul Halsall http://www.fordham.edu/halsall/mod/elizabeth1.asp (accessed 15 June 2013).

13. G. R. Elton, *The Tudor Constitution: Documents and Commentary* (Cambridge: Cambridge University Press, 1982), 274.

14. Lesley B. Cormack, "Britannia Rules the Waves?: Images of Empire in Elizabethan England," in *Literature, Mapping, and the Politics of Space in Early Modern Britain*, eds. Andrew Gordon and Bernhard Klein (Cambridge: Cambridge University Press, 2001), 55.

15. Peter Barber, "England I: Pageantry, Defense, and Government: Maps at Court to 1550," in *Monarchs, Ministers, and Maps: The Emergence of Cartography as a Tool of Government in Early Modern Europe*, ed. David Buisseret (Chicago: University of Chicago Press, 1992), 34.

16. Lesley B. Cormack, *Charting an Empire: Geography at the English Universities, 1580–1620* (Chicago: University of Chicago Press, 1997), 172, n. 31; Barber, "England I," 58; and Victor Morgan, "The Cartographic Image of 'The Country' in Early Modern England," *Transactions of the Royal Historical Society* 29 (1979), 146–49.

17. Marcelo Escolar, "Exploration, Cartography and the Modernization of State Power," in *State/Space: A Reader*, ed. Neil Brenner, Bob Jessop, Martin Jones, Gordon MacLeod (Oxford: Blackwell Publishing, 2003), 33; and Barber, "England II," 68–77.

18. Barber, "England II," 62–64.

19. See Margery Corbett and Ronald Lightbown, *The Comely Frontispiece: The Emblematic Title-page in England 1550–1660* (London: Routledge & Kegan Paul Ltd., 1979), 34; Richard Helgerson, *Forms of Nationhood* (Chicago: University of Chicago Press, 1992), 111–12; and Cormack, "Britannia," 55.

20. Catherine Delano-Smith and Roger J. P. Kain, *English Maps: A History* (Toronto: University of Toronto Press), 260, n. 70. Also see Peter Barber, "Was Elizabeth I Interested in Maps—And Did it Matter?" *Transactions of the Royal Historical Society* 14 (2004): 187.

21. I thank my colleague, Byron Nakamura for the translation of these inscriptions.

> Clemens et Regni moderatrix iusta Britani.
> Hac forma insigni conspicienda nitet.
>
> Tristia dum gentes circum omnes bella fatigant,
> Caecique errores toto grassantur in orbe.
> Pace beas longa, vera et pietate Britannos:
> Iusticia moderans miti sapienter habenas
> Chara domi, celebrisque foris, longaeuaque regnu
> Hic teneas, regno tandem fruitura perenni.
>
> [A gentle governess of Britain rules.
> This mark of distinction shines for all to see.
>
> While sad wars wear down around all the races.
> And blind mistakes wreak havoc upon the entire world.
> You bless the Britains with peace, long and true with your piety:

(ruling with) even handed justice, wisely with a gentle rein,
Held dear at home, while celebrated abroad, and with a long rule you have
this now for such a time, a kingdom everlasting.]

22. Cormack, "Britannia," 55, 57; and Helgerson, *Nationhood*, 111.
23. R. A. Skelton, *Saxton's Survey of England and Wales with a Facsimile of Saxton's Wll Map of 1583* (Amsterdam: Israel, 1974), 10–12. Decades later Saxton's wall map would be the basis for Thomas Jenner's map, called the Quartermaster's map. It was reduced to six sheets in a booklet form in order to be portable, stripped of all its decorative elements, and used by Royal and Parliamentary forces in the midst of Civil War.
24. Arthur Mayger Hind, *Engraving in England in the Sixteenth & Seventeenth Centuries*, vol. 1 (Cambridge: Cambridge University Press, 1952), 167–68.
25. See Edward Heawood, *English County Maps* (London: Royal Geographical Society, 1932), 2, 8–9; Hind, 167–68. These maps are rare and the 1594 version is in a private collection and was printed in the Newberry Library exhibition catalogue, Clark Hulse, *Elizabeth I: Ruler and Legend* (Chicago: University of Illinois Press, 2003).
26. See Edward Lynam, "Woutneel's Map of the British Isles, 1603," *The Geographical Journal* 82, no. 6 (1933): 536–38; Also see Peter Barber, ed., *The Map Book* (New York: Walker and Company, 2005), 142–43.
27. Richard Helgerson "Writing Empire and Nation," in *The Cambridge Companion to English Literature, 1500-1600*, ed. Arthur F. Kinney (Cambridge: Cambridge University Press, 2000), 325.
28. Helgerson, "Writing Empire," 325.
29. Helgerson, "Writing Empire," 325.
30. Helgerson, *Nationhood*, 116.
31. Bernhard Klein, *Maps and the Writing of Space in Early Modern England and Ireland* (Basingstoke, Hampshire: Palgrave, 2001), 107–8.
32. John Brian Harley, "Power and Legitimation in the English Geographical Atlases of the Eighteenth Century," in *Editing Early and Historical Atlases*, ed. Joan Winearls (Toronto: University of Toronto Press, 1995), 165.
33. David Buisseret, "Monarchs, Ministers, and Maps in France before the Accession of Louis XIV," in *Monarchs, Ministers, and Maps: The Emergence of Cartography as a Tool of Government in Early Modern Europe*, ed. David Buisseret (Chicago: University of Chicago Press, 1992), 100–1.
34. Buisseret, "Monarchs," 100–1.
35. Buisseret, "Monarchs," 106. Buisseret suggests that it is unclear if Nicolay's work could have been published if it had been completed.
36. Hilary Ballon, *The Paris of Henri IV* (Cambridge, MA and London: MIT Press, 1990), 213.
37. . . Monique Pelletier, "National and Regional Mapping in France to about 1650," in *History of Cartography: Cartography in the European Renaissance*, Vol. 3, Part 2, ed. David Woodward (Chicago: University of Chicago Press, 2007), 1485.
38. Simon Bouquet and Estienne Pasquier, *Bref et sommaire recueil de ce qui a esté faict, [et] de l'ordre tenüe à la ioyeuse [et] triumphante entrée de tres-puuissant . . . prince Charles IX . . .* (Paris, 1572), 35. See Renaissance Festival Books at the British Library, http://special-1.bl.uk/treasures/festivalbooks/BookDetails.aspx?strFest=0028 (accessed 15 June 2013).
39. Bouquet and Pasquier, *Triumphante Entrée*, 24–35. See Renaissance Festival Books at the British Library, http://special-1.bl.uk/treasures/festivalbooks/BookDetails.aspx?strFest=0028 (accessed 15 June 2013).
40. François de Dainville, "Le Théatre François de M. Bouguereau, 1594. Premier Atlas National de France," *Actes du 85e Congrès national des societies savants, Section de géographie* (Paris, 1961), 22.
41. M. Bouguereau, *Le Théatre Françoys Tour 1594* Amsterdam: Theatrum Orbis Terrarum, LTD., 1966. See Introduction by F. de Dainville, x.
42. Dainville, "Théatre," 27–40; and Ballon, *Paris of Henri IV*, 214.

43. Dainville, "Théatre," 18; Ballon, *Paris of Henri IV*, 214; and Mireille Pastoureau, "Les Atlas imprimés en France avant 1700," *Imago Mundi* 32 (1980): 54.
44. Dainville, "Théatre," 25–26.
45. "Voici le preux Henry des espagnols vaincueur/ Qui deux sceptres puissans maintient d'une main forte/ Cest ce Roy qui pour signe au front la gloire porte/ Le Clemence dans l'ame et la prouesse au cueur. "
46. Maurice Bouguereau, *Le Théâtre français* (Tours, 1594), BN RES L7 2.
47. Petro Plantin, *Gallia* (Antwerp, 1593), BN, Cartes, GE D 17934. The BN record gives the publication location as Antwerp, but Paul de la Houve had a shop in Paris.
48. Ballon, *Paris of Henri IV*, 221. Also see Catherine Bousquet-Bressolier, "Matthäus Merian's 1615 Map of Paris: Its Structure, Decoration and Message," *Imago Mundi* 58 (2006): 52–54.
49. Ballon, *Paris of Henri IV*, 221, 229.
50. See Ann W. Ramsey, "The Ritual Meaning of Henri IV's 1594 Parisian Entry," in *French Ceremonial Entries in the Sixteenth Century: Event, Image, Text* ed. Nicolas Russell and Hélène Visentin (Toronto: University of Toronto Press, 2007), 189–206.
51. Martha Walker Freer, *History of the Reign of Henry IV, kKng of France and Navarre*, vol. II (London: Hurst and Blackett, 1860), 361–62. See also Ramsey, "Parisian Entry," 200.
52. Freer, *Henri IV*, 362.
53. Bousquet-Bressolier, "Map of Paris," 53.
54. See Bousquet-Bressolier, "Map of Paris," 50-54; and Ballon, *Paris of Henri IV*, 220–33.
55. See Bousquet-Bressolier, "Map of Paris," 50-54; and Ballon, *Paris of Henri IV*, 220–33.
56. Ballon, *Paris of Henri IV*, 230.
57. "Nec Pietate Fuit, Belli nec Maior in Armis." See Ballon, *Paris of Henri IV*, 230. Also see Bousquet-Bressolier, "Map of Paris," 67, n. 16 for translation variation.
58. "Sous le regne de ce grand Roy tres clement, tres vaillant, tres juste, Paris est comme souss Auguste fut Rome cu monde l'effroy." See Ballon, *Paris of Henri IV*, 231; and Bousquet-Bressolier, "Map of Paris," 67 n. 17.
59. Ballon, *Paris of Henri IV*, 229.
60. Bousquet-Bressolier, "Map of Paris," 53; and Ballon, *Paris of Henri IV*, 214.
61. Ballon, *Paris of Henri IV*, 227.
62. Ballon, *Paris of Henri IV*, 227.
63. Ballon, *Paris of Henri IV*, 227.
64. Katherine Crawford, *Perilous Performances: Gender and Regency in Early Modern France* (Cambridge: Harvard University Press, 2004), 106–9.
65. Crawford, *Perilous Performances*, 109.
66. Burke, *Fabrication of Louis XIV*, 38–41; Daniel Cady Eaton, *A Handbook of Modern French Sculpture* (New York: Dodd, Mead and Company, 1913), 103–4; and Thomas F. Hedin, "The Petite Commande of 1664: Burlesque in the Gardens of Versailles," *Art Bulletin* 83, no. 4 (2001): 677.
67. Jacques Gomboust, *Le Paris du XVIIe Siècle: Plan Monumental de la Ville de Paris* (Paris: Georges Chamerot, 1890).
68. Marin, *Portrait*, 174.
69. Marin, *Portrait*, 169 and 170.
70. Bousquet-Bressolier, "Map of Paris," 64.
71. Marin, *Portrait*, 172–74.
72. Marin, *Portrait*, 172.
73. See Frank Kitchen, "John Norden (c. 1547–1625): Estate Surveyor, Topographer, County Mapmaker and Devotional Writer," *Imago Mundi* 49 (1997): 43–61.
74. Under James I, though still unsuccessful in *selling* his *Speculum* project, Norden added the surveyorship of the Duchy of Cornwall which accorded him annually 20 marks. Lesley Cormack suggests that Norden's less than successful attempts at patronage to map the counties was due to the government's increased attention to coastal defenses in light of "the imminent threat of Spanish invasion." In addition, Norden's patronage connection to the Earl of Essex who had fallen from favor after his rebellious actions did not help Norden at court. See Cormack, *Charting*, 173.

75. See Kitchen, "John Norden," 46.
76. Edward Holberton, *Poetry and the Cromwellian Protectorate: Culture, Politics, and Institutions* (Oxford: Oxford University Press, 2008), 37.
77. Andrew Gordon, "Performing London: the map and the city in ceremony," in *Literature, Mapping, and the Politics of Space in Early Modern Britain*, ed. Andrew Gordon and Bernhard Klein (Cambridge: Cambridge University Press, 2001), 81. The civic authority of the city is emphasized again in the procession depicted along the bottom of the map as well as the presence not of the royal barge but the "gally fuste" or ceremonial vessel that ferries the Lord Mayor to Westminster to take the oath of office.
78. John Norden, *Civitas Londini* (London, 1600).
79. Kitchen, "John Norden," 58. When Norden died in 1625 he left behind moderate funds and valuable work for the Duchy of Cornwall for which the Prince of Wales' Council was happy to pay a "considerable sum."
80. John Norden, *A Description of the Honor of Windesor* (1607), BL Maps C.27.d.26.
81. John Norden, *The General Perambulation and Deliniation of Cornwall* (1584–1607), BL Maps 196.b.4.
82. Thomas Middleton, *Thomas Middleton: Collected Works* ed. Gary Taylor and John Lavagnino (Oxford: Oxford University Press, 2007), 263.
83. Middleton, *Collected Works*, 264.
84. Stephen Harrison, William Kip, Thomas Dekker, and Ben Jonson, *The Arch's of Triumph* (London, 1604). For image of the Londinium Arch erected for James I's coronation celebration, 15 March 1604. See *Arches of Triumph, Londinium* at the British Museum website. http://www.britishmuseum.org/research/collection_online/collection_object_details/collection_image_gallery.aspx?assetId=54587&objectId=1530077&partId=1 (accessed 12 May 2014).
85. P. D. A. Harvey and H. Thorpe, *The Printed Maps of Warwickshire 1576–1900* (Birmingham: University of Birmingham, 1959), 19; and Andrew Macnair and Tom Williamson, *William Faden and Norfolk's 18th-Century Landscape* (Oxford: Oxbow Books, 2010), 6.
86. John Norden, *London: A guide for cuntrey men* (London: Peter Stent, 1653), BL Maps *3480 (412).
87. Sharpe, "Representations," 865–66.
88. John Ogilby, *The Relation of his Majesties entertainment passing through the City of London* (London, 1661). See British Library site for Renaissance Festival Books, http://special-1.bl.uk/treasures/festivalbooks/BookDetails.aspx?strFest=0248 (accessed 20 November 2012). Ogilby was commissioned to produce the "poetical part" of the coronation celebration ("speeches, emblems, mottoes, and inscriptions"), and he was also awarded by the king the sole right to publish on the event. The first edition was dedicated to the Lord Mayor and Aldermen of London.
89. Ogilby, *Majesties entertainment*, 3.
90. John Ogilby, *The Entertainment of His Most Excellent Majestie Charles II*, 2nd ed. (London, 1662), 1.
91. Ogilby, *Majesties Entertainment*, 10.
92. See Gordon, "Performing London," 84 and 88, n. 39.
93. John Ogilby, *Britannia*, facsimile edition (Amsterdam: Theatrum Orbis Terrarum LTD, 1970), vii.
94. Ogilby, *Britannia*, facsimile edition, v.
95. Sarah Tyacke, *London Map-Sellers 1660–1720: A Collection of Advertisements for Maps Placed in the London Gazette 1668–1719 with biographical notes of the map sellers* (Tring: Map Collector, 1978), 3 and 129; and Ogilby, *Britannia*, facsimile edition, ix–x.
96. Ogilby, *Britannia*, facsimile edition, ix–x. The promised royal patronage of 1000 pounds never materialized but became fairly substantial when the crown did waive the customs duty on the imported paper that Ogilby needed to produce his atlas.
97. John Ogilby, *Britannia* (London, 1675), dedicatory page, NL Case +G 45005.638.
98. Ogilby, *Britannia*, dedicatory page.
99. Ogilby, *Britannia*, dedicatory page.

100. John Ogilby and William Morgan, *A Large and Accurate map of the City of London* (London, 1677), BL Maps C.7.b.4. This is a large map approximately 8.5 feet x 4.5 feet and at a scale of one inch to 100 feet (1:1200).
101. See Richard Pennington, *A Descriptive Catalogue of the Etched Work of Wenceslaus Hollar, 1607–1677* (Cambridge: Cambridge University Press, 1982), 171–73.
102. David I. Bower, "Further Light on Ogilby and Morgan's Map of London (1676)," *Imago Mundi* 65, no. 2 (2013): 280–87.
103. Bower, "Ogilby and Morgan," 284. See Ralph Hyde, *A Large and Accurate Map of the City of London* (London, 1976).
104. Bower, "Ogilby and Morgan," 284.
105. William Morgan, *London Actually Survey'd* (London, 1682). Bower, "Ogilby and Morgan," 284. This is a large map as well approximately 5 feet x 8 feet and at a scale of one inch to 300 feet (1:3600).
106. Morgan, *London.* See image online at Library of Congress URL: http://memory.loc.gov/cgi-bin/query/h?ammem/gmd:@field(NUMBER+@band(g5754l+ct002386)) (accessed 12 May 2014).
107. Robert Morden, Philip Lea, and Christopher Browne, *This Actuale Survey of London, Westminster, & Southwark* (London, 1690). Online Gallery of the British Library has the 1700 edition, http://www.bl.uk/onlinegallery/onlineex/crace/t/zoomify87918.html (accessed 17 May 2014).
108. Burke, *Fabrication of Louis XIV*, 58–59.
109. Some of the material for this section is drawn from my first book, *When France was King of Cartography: The Patronage and Production of Maps in Early Modern France*, 21, 25–32.
110. Petto, *When France was King*, 29–31.
111. Alexis Hubert Jaillot, *Atlas Nouveau* (Paris, 1681), BN, Cartes, GE CC 1002, frontispiece, 1689 edition.
112. Burke, *Fabrication of Louis XIV*, 78.
113. Mircille Pastoureau, *Les Atlas Français XVIe-XVIIe siècles* (Paris: Bibliothèque Nationale, 1984), 233. In recognition of his service to the crown through his map and atlas production, he was awarded, on 20 July 1686, the title of *géographe du roi* (royal geographer) with a yearly stipend of 600 livres. See AN, *Maison du roi*, O[1] 30, f. 247, 20 July 1686.
114. Burke, *Fabrication of Louis XIV*, 135.
115. Alexis Hubert Jaillot, *Atlas Français* (Paris, 1695), BN, Cartes, GE DD 1280, frontispiece.
116. Jaillot, *Atlas Français* (Paris, 1695), BN, Cartes, GE DD 4796(62), dedicatory page. *J'apporte aux pieds de Vostre Majesté le Monde entier comme le Theatre de sa gloire, en luy presentant un nouvel Atlas; où les quatre Parties du Monde, les Empires, leurs Royaumes & leurs Provinces representées dans toute leur étenduë, ne font rien voir de si grand, que le Nom & la Personne Sacrée de Vostre Majesté, qui ont remply toute la terre d'admiration de terreur & de veneration, par la magnificence de son Regne, par ses grandes Conquestes & par la sagesse de sa conduite.* Also as quoted in François Roland, *Alexis-Hubert Jaillot, géographe du roi Louis XIV (1632-1712)* (Besançon: Imprimerie Jacques et Demontrond, 1919), 25, n.1.
117. Jean Baptiste Nolin, *Le Royaume de France* (Paris, 1693), BN, Cartes, GE DD 2987 (386B).
118. *Journal des Scavans* 22, (1695): 141–44.
119. John A. Lynn, *The Wars of Louis XIV, 1667–1714* (London and New York: Longman Press, 1999), 263–65.
120. In the aftermath of the War of Devolution and the Dutch War and in an effort to expand his borders, Louis set up courts to investigate what dependent territories of land that he had already acquired by treaty were still due to him. His actions caused outraged that was in some cases soothed by bribes but ultimately led to the War of the League of Augsburg.
121. Jean Baptiste Nolin, *Les Environs de Paris* (Paris, 1698), BN, Cartes, GE AF Pf 204 (20a-d). See Petto, *When France was King*, 25.
122. Burke, *Fabrication of Louis XIV*, 108–9.

123. James E. McClellan III and François Regourd, *The Colonial Machine: French Science and Overseas Expansion in the Old Regime* (Turnhout: Brepols, 2011), 49.

Chapter Two

Mapping the Land

County and Regional Mapping in England and France

By the second half of the seventeenth century a gradual epistemological shift had begun in the process of making maps.[1] While earlier map makers prided themselves on a meticulous search and detailed analysis of ancient (or former) geographies, itineraries, and other textual sources, that is, the exegetical method, a new generation did not dismiss an exhaustive analysis of sources but championed the use of astronomical observations and measurements by surveyors in the field. In France, Guillaume Delisle (1675–1726) and Jean Baptiste Bourguignon d'Anville (1697–1782), both *géographes de cabinet* (geographers of the study) steeped in the humanistic education and scientific foundation of French geographers, would be prime examples of this new generation that also benefited from the triangulation network being created by members of the Observatory and the Académie, two key institutions of France's Colonial Machine with domestic as well as colonial vision.[2] They were scholar-geographers who were well equipped to analyze field observations and measurements and translate them for cartographic productions and were rewarded by the king with the unique title of *premier géographe du roi* (first royal geographer) with an annual stipend of 1,200 livres.[3]

In England, John Senex (1678–1740), Herman Moll (1644–1732), and Thomas Jefferys (1695–1771), (near) contemporaries of Delisle and d'Anville, "were engravers and map publishers, not scholar-geographers."[4] Among the more well-connected commercial map producers and royal geographers, they did provide maps to the Crown but received no stipend. While they recognized technological advances in surveying and to varying degrees adopted the resulting information, their livelihoods were market driven "rather than by any sustained notion of a 'scientific' geography."[5] Moreover, there was no geodetic project (i.e., no comparable Cassini national map project) being conceived of by London's Royal Society upon which these map makers could draw for regional maps of England nor was there a developing machine or multi-institutional, bureaucratically-driven infrastructure attentive to colonial interests or parallel domestic concerns.

By the eighteenth century the improvement to instrumentation coupled with the appeal to the method of direct observation and measurement in the field provided the foundation for an increasingly specialized group of land surveyors, geographers, and engineers to produce works that participated in the Enlightenment project of authoritative knowledge which could serve the various institutions and corporate bodies of the state from the local to the national level. What distinguished the development of this Enlightenment project in England and France was the bold "visionary leadership" of the Bourbon government to initiate such a project in the 1660s—though not without some lapses in support—to advance it in the first half of the eighteenth century for public work programs and near mid-century to launch a second survey project that would establish cartography as "a routine public activity."[6] In England as Mary Pedley writes: "Neither the Crown nor Parliament was disposed to providing funding for the survey and measure of the realm or to establishing any mapmaking agency."[7] The purpose of this chapter is to follow the production of regional mapping in England and France from the second half of the seventeenth century to the late eighteenth century, and noting, in particular, how map makers benefited from the developing national mapping project in France as well as the entrepreneurial efforts in England to produce county maps prior to the establishment of the Ordnance Survey in the late eighteenth century.

While not dismissing the persuasive power of imagery on maps which was the topic of chapter 1, this chapter focuses on the participation of regional maps in the scientific discourse and Enlightenment ideology of encyclopedism from the late seventeenth into the eighteenth century. It was not a matter that France had better scientists or geographers but Englishmen would argue that the French state supported geographical endeavors more than the English government did.[8] Many French map makers had their roots in a humanistic education and contemporary science as well as connections to the state through the king's ministers (if not the king) and thus enjoyed powerful patronage within the state support for scientific endeavors. They looked to these geographical initiatives and patronage connections to direct and support their work, but in only paralleling the Cassini project these French map makers did face financial challenges partly due to the realities of the commercial trade. In England, map makers may have sought patronage from elites for their entrepreneurial work but "had their roots in the world of trade and looked to the marketplace to support cartographic endeavors."[9] Only with the establishment of the Ordnance Survey in the late eighteenth century and its regularization in the nineteenth century did the cartographic practice of regional mapping become part of larger institutional structure.

OGILBY'S COUNTY MAPS

In England regional mapping in the seventeenth century was drawn from the earlier *Atlas of the Counties of England and Wales* (1579) by Saxton or the works of his successors, such as Norden and John Speed (1552–1629). Saxton's work was "shamelessly copied" with little added information and without new surveys being completed.[10] As a consequence, Ogilby's plan to survey counties anew (by the road traverse method) as part of his grand project was a welcomed endeavor. His patrons may have come to know his works from his early publications of translations, his participation in the planning (and publication) of the coronation celebration for Charles II, or when Ogilby was appointed one of the "sworn viewers or surveyors" for the rebuilding plans after the Great Fire of London.[11] Given the challenges that post-Saxton projects encountered, it remained to be seen if such a useful project would garner sufficient patronage.

By the 1670s, as part of his larger atlas project, Kent and Middlesex were being engraved and a survey of Essex was underway by Ogilby's assistant, Gregory King (1648–1712).[12] In 1672, the survey in part financed by Gilbert Sheldon (1598–1677), Archbishop of Canterbury, was published in the form of a map, *A New Map of Kent Actually Survey'd and Deleniated by his majesties cosmographer John Ogilby, esq.*[13] Sheldon, the former bishop of London, was known for his charitable support and patronage of learning and helped bring to light this survey by Ogilby. Sheldon's coat of arms adorned the map alongside the dedicatory cartouche to this "munificent patron" and member of the king's Privy Council. The Archbishop's arms were joined by the armorial bearing of the King atop the title cartouche. Shortly after Ogilby's death (1676), his successor William Morgan published the survey of Middlesex, made possible by the patronage of Sir Thomas Wolstenholme (c. 1622–1691), a member of the gentry class and a prominent baronet of Edmonton parish.[14] The map, produced by survey, reflected the unique presence in this county of the crown and the City and so carried the prominent coat of arms of the city of London atop the title cartouche with the royal arms just above it, and opposite, Wolstenholme's arms appeared with the dedicatory cartouche. Similarly, Morgan was able to publish the survey of Essex: *Essex actually surveyed with the several roads from London etc. exactly measured at three mile in an inch by John Ogilby and William Morgan his majesties cosmographers 1678.*[15] The work was dedicated to one of Ogilby's early patrons, Arthur Capell (c. 1631–1683), Earl of Essex and a member of Charles II's Privy Council.[16] The arms of Capell and the king adorn the map while the scale cartouche, similar to the other county maps, displayed playful cherubs with surveying paraphernalia.

These maps showed the counties divided into hundreds with some topographical features such as roads, parks, elevations, churches, or windmills, and while Sheldon, Wolstenholme, and Capell had supported the efforts to bring these maps to the public, Ogilby's plan for a new survey of the counties of England a century after Saxton's work was financially untenable. Despite the patronage from the crown, the elite nobility, and even the gentry, it was insufficient to move ahead with the larger atlas project or even more individual county maps. These three maps drawn from surveys carrying the royal arms as well as those of its other patron saw the light of day because of the individual support of these elites, but with no sustained effort to continue the surveying project and no central government or even organized local directing agency, a new county atlas would not be achieved. Local elites were more likely to call upon a surveyor for a more immediately and personally useful estate mapping plan than any larger project[17] (see below).

EARLY REGIONAL MAPPING IN FRANCE

As reported earlier (see chapter 1), there was no survey-type Saxton project in sixteenth-century France. Early regional mapping leading to printed maps could be found at the more local level such as diocesan mapping by local ecclesiastics.[18] The importance or power of maps, however, was quite well understood. Jean Jubrien (1569–1641), a local map maker, dedicated his 1623 map of the diocese of Reims to the new archbishop, Gabriel de Sainte-Marie (1623–1629), now duc de Reims and premier peer of France.[19] Perhaps in recognition of the Archbishop's prominent position in France, Jubrien suggested in his dedication (on the map) that just as a captain leading an army would find such a description useful, the Church could be equally served in its spiritual duties, as it needed to be prepared to do battle with the devil, the world, and the flesh. At a single glance, a map could demonstrate the potential gravity of a situation.[20] Over fifty years later, Père Augustin Lubin (1624–1695), *géographe du roi* (royal geographer), remarked that newly consecrated bishops ought to seek exact maps of their ecclesiastical domains in order to know the face of their flocks, as well as to assist them in putting their dioceses in order.[21]

Until the late seventeenth century much of mapping whether regional or not was drawn from mémoires and travelers' itineraries. Nicolas Sanson (1600–1667), the former military engineer turned map maker and among the first to produce French originals and considered the father of French cartography, was a *géographe du cabinet* (geographer of the study). Having come to the attention of Louis XIII's chief minister, Armand du Plessis, Cardinal

Richelieu (d. 1642), Sanson was soon tutoring the king and others in geography.[22] Sanson's efforts to organize and publish an atlas befitting his young student provided him with great governmental support, but Sanson was of the old epistemological school and much of his work was drawn from a compilation and detailed analysis of his sources. He was praised in the field for his tables which organized geographical information and assisted others in the drawing of maps. By his death in 1667, he had produced one hundred diocesan maps and left many in manuscript form. On account of his reputation in the field and his patronage connections, Jean-Baptiste Colbert, one of his patrons, had appointed Sanson to evaluate the reports coming in from the provinces for geographical information, but his unexpected death would leave a new generation to lead the field.

About a decade before Ogilby published his first county map, Colbert, Louis XIV's finance minister and the architect of Bourbon *gloire*, enlisted the services of the provincial *intendants* to advance (among other things) the royal government's geographical knowledge of its provinces.[23] In 1663 Colbert sent his request for information, *Instructions pour les maîtres des requêtes, commissaires départis dans les provinces.*[24] As the reports made their way back to the court, it was clear to Colbert that in terms of geographical information much work was needed, and map making and the pursuit of geographical knowledge became one of the primary concerns of the soon to be established (1666) Académie.[25]

Within the last two decades of the seventeenth century, the likes of Jean Picard, Philippe de la Hire and Jean Dominique Cassini (lured from Italy in 1669 with the promise of royal patronage), revealed just how much field work was necessary as they presented (in 1682) to Louis XIV a Sanson map of France with a new outline of the coastlines based on their astronomical observations (and mathematical measurements). The King was shocked that his mathematicians and astronomers had dispossessed him of more land than any of his successful military campaigns had subjugated for him.[26] That map, published in 1693, made public what value there was to the observations and measurements of the astronomers and mathematicians of the Paris Observatory and Académie. While this project to create a triangulation network to span the country and aid in the production of a national map survey suffered from inconsistent royal support into the early eighteenth century, it influenced the pursuit of empirical information and the production of maps as instruments of power. In contrast to chapter 1 that looked at the representational frame with decorative elements, in this chapter the entire map is viewed as a scientific production. The map still participates in the discourse of power, but, in place of or in addition to the entertaining bits and pieces, the message touted on (or with) the map is that it was born out of a methodology, that is, it is a scientific production, and it was to serve the interests of power.

Three decades after Colbert's initial call for information from the provinces, Paul de Beauvilliers, duc de Saint-Aignan (1648–1714), governor of the Dauphin, ordered the intendants to write reports concerning their regional jurisdictions for the instructions of Louis, duc de Bourgogne (1682–1712), second in line to the throne after his father, the Grand Dauphin (d. 1711). The responses to the questionnaires sent out by Beauvilliers included geographical information and resulted in the manuscript collection a few years later from 1697, *Mémoires des Intendants sur létat des Généralités*.[27] Such a wealth of information including names and descriptions of cities, villages, dioceses, municipal officials, royal officials, population figures, agricultural information, land use, public work projects including the state of roads, billeting (i.e., military) capabilities, prominent buildings or sights, among many other elements would certainly be useful to a future monarch and assist in the compilation of maps, but also indicative of lacunae in geographical knowledge. These court-centered initiatives born out of the program of Bourbon *gloire* also further supported the concept that geographical knowledge was a significant element of the promulgation and execution of sovereignty.[28] These efforts were the domestic arm of the developing Colonial Machine whose infrastructure served the French state at home. In the 1690s with war, economic crises, and famine, all administrative offices serving the state would benefit from the process of compiling these mémoires as well as the cartographic exercises launched by the earlier Colbert request and this new proposal by Beauvilliers.

Among these surveying efforts were those which appeared in the province of Languedoc, a hotbed of Protestantism before and after the Revocation of the Edict of Nantes (1685). For example, Hubert Gautier (1660–1737) from Nismes in the province of Languedoc left medicine to pursue a career in mathematics, became a royal engineer, and served as an engineer in the province (beginning 1689) for nearly three decades. He converted to Catholicism in 1689 in the presence of the bishop of Nismes, Esprit Fléchier (1632–1710), who was not only known for his great eloquence, but his charity and piety in serving the members of his flock. Gautier's role as architect, royal engineer, and inspector of roads made him particularly qualified to make maps of the dioceses of Languedoc. Gautier hoped to fill in the gaps left by the seventeenth-century diocesan maps of Sanson. The intendant of Languedoc (since 1685), Nicolas de Lamoignan de Basville (1648–1724)—a royal agent who gathered around himself his own useful clients—put Gautier to work on the survey and construction of the roads of the province.[29] In response to Beauvilliers' request, Intendant Basville contributed his *Mémoire* for Languedoc and although neither the manuscript nor printed work included a map, Basville's efforts did lead to several maps.

Among Gautier's first surveying efforts was the diocese of Toulouse that he drew and dedicated (1695) to Jean Baptiste Michel Colbert de Saint

Pouange (1640–1710), Archbishop of Toulouse.[30] The printed map, published by Jaillot, *Le Diocese de Toulouse*, and also dedicated to Archbishop Colbert, used Gautier's survey though did not copy it. Gautier's name did not appear on the map nor did Jaillot indicate that it was from a survey, "*levée en scene*."[31] The appearance of parcels of land on the map divided into dioceses and archdeaconries gave it an orderly appearance and perhaps the impression that it was surveyed, but Gautier's map did not include this engraved embellishment. On Jaillot's map, the title cartouche carried the prominent arms of the Archbishop, a cousin of the great Colbert, but did not include the dedication on Gautier's map in which he reflected on Colbert's great service to the church and even his service to the state. Around the cartouche on Jaillot's map were the arms of the principal cities of the diocese over which the Archbishop had ecclesiastical power and political influence. Gautier's map included no imagery. It is unclear how Jaillot obtained Gautier's map. Gautier could have sold it to Jaillot and given up his right to offer a dedication on the map, or Colbert could have asked Jaillot to publish the map and agreed to the dedication by Jaillot. In either case, this royal engineer, Gautier, directed in his surveying efforts by the king's intendant, Basville, was able to serve the interests of the central state, the secular responsibilities of the ecclesiastical leadership, and the Church, and perhaps also provide a visual document that could accompany Basville's *Mémoire*.

A few years later, Nolin published a map of the diocese of Nismes drawn from Gautier's survey, *Le Diocese de l'Evesché de Nismes*.[32] On this map, Gautier's name did appear in the title as well as an indication that the map was drawn from a survey, "*dressé nouvellement sur les lieux*" and accomplished with an instrument of his [Gautier's] own fashioning. Whether he used triangulation to create the map is unclear, but it is unlikely. In 1698, just a year after Basville submitted his manuscript on Languedoc, Nolin engraved and published this more highly decorated map and dedicated it to Bishop Fléchier, a friend and ally of Basville.[33] The title cartouche with the dedication to the bishop carried his coat of arms at the top, and below the title cartouche was a decorative drape on which the plan of the city of Nismes was depicted with its seal at the top, along the edges were the seals of the five other principal cities of this diocese, and to the side a list of prominent places in the city. Nolin, an engraver and map seller, filled the outer borders of the map with famous public projects some of which were described in Basville's *Mémoire* such as the amphitheater of Nismes, the Pont du Gard, the Temple of Diane, the Maison Carrée, as well as a list of symbols for finding walled cities, parishes, hamlets, churches, and chateaus. As Nolin reported on his map, the work was ordered by Fléchier and made under his care.[34] While the work not only highlighted information in the *Mémoire* for Languedoc, it was a means of ordering the local administration in the midst of Protestant tensions, and in light of the creation of the new diocese of nearby Alais (in 1694), a visual document of ecclesiastical power.[35]

In 1697, the date of Basville's *Mémoire*, Nolin also engraved and published a small-scale map for the province (*Gouvernement Général*) of Languedoc which he dedicated to Cardinal de Bonzi (1631–1703), Archbishop of Narbonne and President of the Estates of Languedoc, who had been on account of ill health losing power in the province and was slowly being discredited by the younger and politically savvy Basville.[36] While the title cartouche included the coat of arms of the major cities of the province, the dedicatory cartouche included Bonzi's ecclesiastical accoutrements as well as some mediocre engravings of symbols of the church and abundance. The legend for the map included a dizzying array of symbols for archbishoprics, abbeys, sea ports, forests, salt storehouses, universities, the seat of parlement, etc. On the map, Nolin noted that the geographical information for this work was drawn from the *Description du Royaume de France* (1693) by Sr. Tillemon, a.k.a. Jean Nicolas de la Tralage (d. 1699), a useful source of information for Nolin. In addition, Nolin commented on the location of the principal cities whose latitudes and longitudes "conformed" to the observations and measurements of the members of the Académie and concluded that "in a word, this map is in many ways more exact that any which have appeared until now."[37] Nolin was able to print such a statement thanks to the information provided by Tillemon, but after his death in 1699, Nolin had to find other means to keep his work up to date and was soon embroiled in a plagiarism case with the Guillaume Delisle.[38] These works not only reflect the secular and ecclesiastical participants upholding the interests of the state in provincial administration, but also the effort by map makers, in particular by Nolin, to connect their work explicitly to the contemporary scientific discourse. It may have been an empty rhetoric aimed only to attract a consumer to his map or to alert a potential patron that he was keeping abreast of developments in cartography.

In addition to the surveying being conducted in Languedoc, understandably the region around Paris received a great deal of attention as founding members of the Académie, Jean Picard (1620–1682), Gilles Personne de Roberval (1602–1675), and Jean Dominique Cassini (1625–1712), began conducting observations, making measurements, and checking instruments for what would become the national mapping project of France. The engineer, David du Vivier benefitted from their work and had his instruments and measurements evaluated by these astronomers and mathematicians.[39] The resulting map from these initial efforts of the members of the Académie was a grand nine-sheet map of the region around the city of Paris printed in 1678 at a scale of less than one inch to a 1.5 miles and carrying the coat arms of the king which covered nearly half of one of the sheets.[40] Adorning the cartouche of the scale of measures was the arms of the city of Paris, but in place of its motto, *fluctuat nec mergitur* (it floats and does not sink), was *felicitas populi* (happiness of the people) which is reminiscent of what Nolin wrote on his map of Paris (1698) about the fortunate and happy Parisian subjects of Louis XIV(see chapter 1). Colbert's coat of arms, accompanied by the watchful eye—symbolic of his information network—and images of his finance ministry, was placed above the legend which was also adorned with the instruments of drawing and measuring. Any viewer using the map would be able to identify cities, towns, hamlets, farms, windmills, watermills, chapels, parish churches, etc., and notably on account of the authors of the map, the Paris Observatory. There are no roads on this map, but the rivers and the streams (and the numerous watermills) give the impression that internal movement and transport was very much influenced by these natural byways.

Among the early maps to draw from this work was that by Albert Jouvin de Rochefort (c. 1640–c. 1710), a cartographer and financial officer for the crown, who dedicated his map of the Archdiocese of Paris to François de Harlay (d. 1695), archbishop of Paris, who enjoyed the favor of the king and spoke out on ecclesiastical matters.[41] In the 1690s Jaillot published a two-sheet map, *Généralité de Paris*, which he dedicated to the intendant, Jean Jacques Charron whose coat of arms adorned the title cartouche with the images of Minerva and Mercury.[42] This small scale map may not have provided much detail, but the viewer was able to follow the boundaries between the various administrative units (généralité) of Paris, Rouen, Orléans, etc. In fact, Jaillot's map accompanied the *Mémoire* for Paris written by Charron's successor, Jean Phélypeaux (1646–1711), a few years later.[43]

In similar fashion to these works, Jacques Benigne Bossuet, the influential bishop of Meaux, ordered François Chevalier, mathematician and member of the Académie, to draw a map of the region. In 1698, Jaillot published the map of the diocese of Meaux and, as a very humble and very obedient servant, dedicated it to the bishop.[44] Engraved by Louis Cordier, the map title cartouche carried the dedication, the arms of Bossuet, and the symbols of his ecclesiastical station. The map included plans of the bishop's palace at Germigy and the city of Meaux with notations for locations such as hospitals, religious houses, or city gates, and notations for market days that appeared next to principal locations. The cartouche with the scale of measures also provided the symbols for priories, towns, chateaus, farms, etc. as well as the ecclesiastical divisions within Meaux. The map displayed the various forested areas, the fields, roads, and rivers. Surveyed on location, the map provided great detail and was so well received that Cardinal de Noailles sought a map of the Archdiocese of Paris and set the geographer Jean Besson to work on it.[45]

While these on site works by Gautier, Vivier, and Chevalier directly or indirectly commanded by the government were more advanced than those earlier seventeenth-century maps drawn from itineraries, the maps produced from them were often publications from the businesses of map publishers or engravers. As business owners, Jaillot and Nolin (both of whom carried the title of *géographe du roi*, royal geographer) were entrepreneurial in their efforts to provide maps for influential patrons but, for lack of training were only able to participate tangentially (e.g., purchase a manuscript map) in the new epistemology which saw the influence of a scientific methodology in map making. The cartographic activity, however, within the Delisle workshop reflected this new epistemology. He emphasized that his works were different than those that had come before (see chapter 4). While, thanks to his father, Claude, he appreciated the importance of a careful study of textual sources, he was also well-versed, thanks to his connections to the Académie, with the importance of astronomy and mathematics to the cartographic enterprise. It was this approach that would earn Delisle the unique title of *premier géographe du roi* (first geographer to the king) in 1718.

Much of the cartographic enterprise was state supported and directed, and with the work of Delisle, the state and the Paris map trade were served by a geographer with impressive royal patronage, connections to the Académie, and a concern to produce accurate maps drawn from observations and measurements.[46] Among his works were several regional maps of France including ecclesiastical diocesan maps. In 1708 Delisle published a map of the Diocese of Béziers drawn by Gautier, but corrected by Delisle—an ability that neither Jaillot (map publisher) nor Nolin (an engraver) had.[47] Dedicated to and carrying the coat of arms of (Louis-) Charles des Alris de Rousset (1662–1744), bishop of Béziers, the work was corrected in accordance with the observations of the members of the Académie and, uniquely, carried a privilege of twenty years dated from May 1708.[48] Unlike the previous Jaillot and Nolin maps, Delisle's included a finding aid. Along the left side was an alphabetical list of locations and coordinates to aid the reader in finding places on the map. To the upper right side was the plan of the city of Béziers similarly with a table of thirty-six locations noted on the plan. Below was a more detailed list of symbols used on the map, such as roads, bridges, abbeys, chateaus, archdioceses, dioceses, royal seats, ports, etc. Along the bottom was a description of unique things in the area, such as marble mountains and mineral springs, and there was a much clearer indication on Delisle's map of the mountainous terrain and other areas of elevation.

The following year, in 1709, Delisle published a map of the Diocese of Senlis, one of the outstanding topographical maps prior to the publications from the work of the Cassini project.[49] The work was based on a map drawn on site by M. Parent, a curate of d'Aumont (in the region) and, like the Béziers map, corrected by the astronomical observations and geometric operations of the members of the Académie. Delisle dedicated the map to Jean François de Chamillard, bishop of Senlis, whose ecclesiastical symbols and arms adorned the map. Plans of the city of Senlis and Crespy with their principal sites were included as insets. Again, uniquely, the map carried a privilege dated from 28 August 1709 and the finding aid for the local dioceses along the left and right sides of the map. On account of the quality of such work, the reputation of the Delisle workshop, his patronage connections, and his map making methodology, Delisle was approached by the revenue officials of Burgundy to embark on a much larger project than these diocesan maps.[50]

In the midst of the famines of 1708 and 1709, the provincial officials of Burgundy, perhaps requiring greater geographical knowledge to mount a relief effort, approached Delisle to produce a map of the Duchy of Burgundy, an area substantially larger than the diocese of Senlis. Delisle's reputation as a talented geographer could not really escape the fact that no survey had been completed and that information would have to be compiled from rather disparate sources. In 1708, these revenue officials negotiated with Delisle to make a map of the Duchy of Burgundy based on—as per Delisle's method—maps, interviews, observations, measurements, and mémoires. Maps provided by locals such as Philippe Languet de Sivry, the civil lieutenant of d'Arnay-le-Duc, and others from local surveyors such as Canut assisted Delisle's progress along with observations from the Académie.[51] Others sent information to Delisle which can be found in his correspondence. Once the map was completed, Delisle sent a copy to the president of the Estates in order for him to make any necessary corrections. The resulting two-sheet map, published in 1709, was the best contemporary map of the area. It was engraved by Desrosiers and by Nicolas Guerard, whose title cartouche—designed by the revenue officials—was adorned with the coat of arms of the hereditary governors of Burgundy, the Condé family.[52] In the midst of his work to complete the map, Delisle wrote to the officials of Burgundy and even to the Duke (de Bourbon), seeking a more appropriate financial arrangement.[53] He expressed his frustration that to accomplish such a work required a greater financial outlay than originally thought. He had, after all, been working for an entire year solely on this project whose expense would nearly equal the contracted payment.

Despite its beauty and acclaim for quality and the reputation of the geographer, such a map even directly benefitting from the crown's support of the efforts of academicians and a parallel effort by provincial officials was financially draining without a greater infrastructure in place. Within the decade (1716), the regency government under Philippe d'Orléans (a patron of the Delisle family) had directed the Académie to initiate a new effort to gather information about the provinces from the intendants.[54] Perhaps in light of these more recent efforts, Delisle's reputation, and his experience with the project for Burgundy, in January 1722, the Estates General of the province of Languedoc decided to undertake a project for an atlas of Languedoc.[55] The project "was to be an achievement of the province, which thus had the opportunity to show 'its zeal in the service of the king and in the good of the public,' and to boast of being an example to all the other provinces."[56] Delisle began negotiations with René-François de Beauvau (1664–1739), archbishop of Narbonne and the president of the Estates. Delisle sketched out a plan that he hoped would bring together the earlier work of the academicians from Paris on the length of the meridian, the particular maps of local geographers, and the, then current, efforts by the members from the Society of Sciences at Montpellier to extend the triangulation network through the province. In effect, Delisle created a mini-engineering project (à la Cassini), but in incorporating the Science Society, he not only conferred greater credibility at the local level, but also gave evidence of the "symbolic filial link" between the Society at Montpellier and the Académie at Paris.[57]

Delisle's untimely death (in 1726) and his widow's inability to strike a deal with the province's syndic, Montferrier, was a blow to the progress of the atlas project.[58] In spite of the "filial link," tensions existed between the center and the periphery as the Society (periphery) insisted that all fieldwork would be assessed first locally before being sent to Paris.[59] With fieldwork only beginning in 1730 and now apparently only local oversight and impetus driving the project, progress was slow. It would ultimately be taken over by the Cassini second national survey project (see below). While Delisle's early negotiations had been a start for a provincial mapping program and clearly mirrored the national mapping project, the plan proved too ambitious for the present infrastructure. What was needed or demanded by the crown was a centrally-directed program rather than individual projects—no matter the talents involved—between the provinces and perhaps Parisian map makers. English surveyors in the eighteenth century had no center to either benefit from or to challenge for authorship and faced a daunting task as they set out to make surveys or were commissioned to make surveys of counties five to ten times smaller than the Duchy of Burgundy or province of Languedoc.

EARLY COUNTY-SURVEYING PROJECTS IN ENGLAND

At the end of the seventeenth century when London's commercial mapping trade was still quite saturated with small-scale county maps, the earliest effort to produce a large scale county survey was begun by Joel Gascoyne (1650–c. 1704), for the county of Cornwall.[60] Gascoyne, a surveyor, engraver, chart maker, and map maker, carried out a trigonometric survey (i.e., triangulation) over the course of six years (from 1693 to 1699) and printed a broadsheet to announce the publication of his map to those who had already offered some support and to new potential subscribers.[61] This kind of map activity ushered in what historians of geography refer to as the second phase of English county mapping, a period distinguished by private enterprise (or initiative).[62] These maps were produced through subscription efforts and by local patronage, but, in contrast to France, this activity did not parallel a national program. Moreover these productions appeared in a market already saturated with small scale maps and atlases of the English counties.

Gascoyne, a member of the Thames School of chart making had abandoned this earlier work around 1689 perhaps because of the death of his, then patron, the Duke of Albemarle, Governor of Jamaica.[63] His former chart making and estate mapping, however, led to commissions to survey the estates of Lord Bath in Cornwall and the estate of the Earl of Radnor.[64] These smaller projects led Gascoyne to undertake the large county map of Cornwall which he dedicated to Charles Bodville, Earl of Radnor, and whose coat of arms adorned the dedicatory cartouche. Upon completion of the surveying effort, which took about six years, Gascoyne, in a broadsheet, addressed (27 March 1699) the nobility and gentry of the county of Cornwall not only to promote his work but to seek "encouragement" from the county's elites to alleviate the great expense of publication.[65] Delisle's effort to publicize his expense to produce his map of Burgundy involved writing a private letter to the officials and making an appeal for patronage to the duke, a much less vocal and more traditional method. The earl was certainly one of Gascoyne's patrons but would not solely finance the project.

Gascoyne reported that he had "with indefatigable pains" and "almost insuperable difficulties" completed the map of Cornwall with "the names, and principal seats, and families of the county, which I perswade my self, will in some measure, contribute to transmit them to posterity."[66] After all, here was a work in which county elites could view their estates and properties in relation to their neighbors, and with their principal seats marked on the map, anyone who viewed it would see their social prominence. County maps, unlike the more personal estate plans, afforded many more eyes to see the lay of the land and the power within it. Gascoyne wanted to assure his future patrons and/or subscribers that he had included many locations—though not too many so as not to burden the eye—with great accuracy and few errors.[67] In an effort to assure the accuracy of his rendering, he exposed "the proof to publick view, which may be seen at the Grand Jury Chamber." True to his charting roots, he also paid attention to the coastlines, for instance, harbors, capes, and the like. And finally having come to the end of his labors, he hoped for financial support in the form of a subscription: "I humbly present it to your Honours favourable acceptance; and with an humble confidence, assuring my self, your goodness and candor, will give encouragement suitable to such laborious, publick undertakings."[68]

His appeal for encouragement was based on his effort to acquire the funds for the printing of the work. The work in twelve sheets, *A Map of County Cornwall Newly Surveyed by Joel Gascoyne*, with a separate sheet for the names of prominent cities and towns and a list of tenants, sold in London at several shops including that of Richard Mount, Philip Lea, and John Thornton, the last a fellow member of the Thames School and with whom Gascoyne had been apprenticed.[69] It also sold outside of London in Exeter and Launceston in county Cornwall and Plymouth in county Devon. Adorned with a scene of surveying in action showing the circumferator and the theodolite, it was a seemingly successful calling card for future work, but at this early stage progress was still tenuous. In the following year Gascoyne had proposed a survey of the nearby county of Devon and sought subscriptions for the work, but he was unsuccessful.[70] Gascoyne's second project might have failed because of the death of Lord Bath (d. 1701) and shortly thereafter, his successor robbing Gascoyne of the foundational patronage that might have allowed him to start the work. There simply was not enough interest, no directing institution, no parallel national program to exploit or emulate for Gascoyne to risk the financial outlay for a survey.

A few decades later John Senex proposed a county survey of Surrey. In the eighteenth century, Surrey just south of Middlesex County became a convenient location for London elites to establish or purchase estates. After the success of his *New and General Atlas* (1721) dedicated to the Prince of Wales (future George II), Senex had demonstrated the importance of geographical works to those with social prominence as three quarters of his subscribers to this atlas were from the landed elite.[71] Unlike Gascoyne, Senex maintained a successful business on Fleet Street in London where he sold maps, globes, atlases, and geographical works as well as partnering with other publishers to produce geographical works. As a geographer, engraver, globe maker and member of the Royal Society, he championed the use of observations and measurements from the Royal Society and Académie, so it is not surprising that he embarked (as early as 1721) on a project to survey the county of Surrey and to produce a map (in four sheets) at a scale of one inch to a mile.[72]

According to Laurence Worms, his novel method of survey would involve "such gentleman [sic] as are furnish'd with proper instruments" who would calculate their bearings by multiple observations: "A ball of fire will be thrown up from the top of Box-Hill at half an hour past eight every evening for a fortnight; and rockets will be let off from proper eminences near Godalming, Chertsey, and Westram; the first half a quarter of an hour after the ball, and the rest at a like interval from each other."[73] It is unclear if the entire survey of Surrey was accomplished in this fashion and how active Senex was in the process. This endeavor, however, must have been quite a show, and despite participation in at least three other county map projects, he did not duplicate this unique event elsewhere.[74]

In 1729, upon completion of his work, it was sold at his London establishment as well as at the nearby London shops of Thomas Bowles and E. Cushee. Subscribers to the work were rewarded with a high quality map depicting locations of principal seats and the names of the proprietors, prominent locations, roads, rivers, parks, churches, fields, and so on. While the map carried no specific dedication, around the border of the map were 156 shields, of which surprisingly, about only 25 percent included coats of arms. These active coats of arms indicate the subscribers to the piece whose principal seats could be seen on the map.[75] The few number of arms depicted gives an indication that he was not able to get enough "encouragement" to make it financially viable.[76] This Surrey project "wounded" him, that is, the financial outlay and labor was too great for the return in the trade and could not be sustained.[77] Without more social potency these individual initiatives were likely to flounder. The audience or consumer group to draw on was simply not large enough, and there was no state office or engineering corps that might benefit from using these maps in their official duties or in taking up the task to produce these maps for the state.

Similarly, Henry Beighton (1687–1743), land surveyor, mathematician, editor of the *Ladies' Diary* (since 1713) and fellow member of the Royal Society since 1720, proposed a map of his home county, Warwickshire, in 1722. As a land surveyor Beighton had witnessed the changes in the use of land from large open fields to smaller individual parcels and the loss of agricultural land to coal mining.[78] An earlier attempt in 1711 to propose a survey never got off the ground, but Beighton remained critical of the current maps that continued to draw from the sixteenth-century works of Saxton or Norden. A proper survey by triangulation was necessary, and he intended to complete it. Beighton returned to his project when he became involved in drawing new illustrations and maps for a new edition of *Antiquities of Warwickshire.*[79] He proposed to publish a large and correct map of Warwickshire, but he needed support. He looked to the marketplace: "In order to facilitate the great expence [sic] of such survey, which can't well be done without some assurance of the Nobility, Clergy, and Gentry's favourable concurrence, it is proposed by subscription."[80] The seats of the nobility and gentry would be noted on the map and their arms would adorn the borders of the map with an index so that one might easily connect the arms with the location of the principal seat.

Seeking subscribers, Beighton advertised his map several times in the *Ladies' Diary* between 1723 and 1729 even reporting in the 1723 edition on the quality of contemporary surveyors as "Lame, Deficient or Trifling."[81] Upon completion of the surveying in late 1725, he found the number of subscribers insufficient for the necessary financial outlay. In 1727 still hoping for more subscribers he offered that a specimen could be viewed at the offices of Mr. Elisha Kirkall (who engraved the map) and Mr. John Senex. Finally the map began to be delivered in summer 1728. In the *Ladies' Diary* Beighton (again) urged "any of the nobility and gentry who have not already subscribed for their arms to be engraved on the map to do so without further delay," and, perhaps in an effort to shame local elites into subscribing, he reported in the *Ladies' Diary* of 1729 that fifty-three noblemen and gentlemen had not (yet) subscribed to the work. [82] The print run was most certainly less than 300 and probably did not exceed 250. Beighton may have reached about 85 percent of the numbers he hoped to achieve, but while different extant copies of the map showed some variation, what is clear is that the blank shields displayed along the side borders of the map were most likely intended to be *with design* and the many blank shields attest to the lack of subscribers or Beighton's optimism for interest in the map. [83]

The market was still satiated with county maps in individual sheets or compiled into atlases. For example, Herman Moll's *New Description of England and Wales* (1724), a work of more than four hundred pages with *new* county maps adorned with images of local British and Roman coins and antiquities boasted "fair Maps . . . much rectified and improved, by the careful Perusal of several valuable and exact Surveys."[84] These county maps with accompanying descriptions provided the same local gentleman with an armchair delight of all the counties. The financial outlay, the time-consuming surveying, and the requirement for substantial local interest for any county mapping project to be successful doomed these early efforts as there simply was no infrastructure in place and a *market-driven* project could not be sustained. In England, twenty-five years later efforts to support county surveys would emerge again, and in France, regional mapping efforts paralleling the work of the national map survey would intersect in the second half of the eighteenth century providing the infrastructure for these local projects.

RAISING ACCURATE REGIONAL MAPS BY METHOD, EIGHTEENTH-CENTURY FRANCE

In an effort to address the expense to complete a survey on site, the French mathematician, François Chevalier addressed the use of the local infrastructure already in existence. He had been asked by Bishop Bossuet, in the late seventeenth century to draw a map of the diocese of Meaux, so he was aware of the technology required. In 1700, Cassini with his son, Jacques, returned (after nearly two decades of little work in the field) to the triangulation network along the meridian of Paris, but once again war (War of Spanish Succession, 1701–1713) and its financial and political concerns drew attentions elsewhere. In the early eighteenth century perhaps in an effort to continue making progress Chevalier presented to the Académie an inexpensive method of raising accurate maps on location using amateurs. [85] Under professional guidance and supplied with base maps, amateurs would fill in the required information. Each base map, provided by the professional, consisted of eight concentric circles divided into eighths with the largest radius measuring two leagues. (That means the area to be described had to be smaller than four leagues.) With clear notations of the location of the rising and setting sun along the east-west orientation, the amateur filled in the base map with the desired information and then returned his map and accompanying report to the professional who compiled the information with measurements and astronomical observations already recorded to create a map of the area.[86]

Shortly after Chevalier presented his map making project and during another decade of lack of government support for and inactivity of the initial Cassini project to establish the meridian of Paris (finally completed in 1718), Jean Baptiste Bourguignon d'Anville put into action Chevalier's method for regional mapping.[87] At the same time that Delisle was in negotiations with the Estates of Languedoc, d'Anville commenced a plan to map the diocese of Lisieux for Henri Ignace de Brancas (1687–1760), the bishop and comte de Liseux, a "man of science" himself and an advocate for the improvement of geographical knowledge and the use of astronomy.[88] Unlike Beighton who was at the mercy of Warwickshire elites to financially buoy his surveying and publication efforts, d'Anville, who enjoyed the patronage of the powerful Orléans family and a well-respected position in the sphere of geography, seemingly had a captive audience to conduct his work and to finance it.

A half century or more after similar comments by the map maker, Jubrien, and the royal geographer, Père Lubin, d'Anville suggested that such maps would assist bishops in the administrative function—both civil and ecclesiastical—of their states such as the care of bridges and roads as well as the pastoral duty to provide the sacraments to distant places.[89] This diocese, located in the ecclesiastical province of Rouen (Normandy), had been enjoying increased economic opportunity due to an expanding manufacturing base, and as a result of the growing financial benefits in Normandy, many journeyed from the south and the west to seek work. With an increase in the "faithful," local and provincial governments increasingly found themselves in the business of poor relief, one of the concerns defining new state activity in the second quarter of the eighteenth century.[90] Mapping efforts such as this one by d'Anville, who was solicited by the bishop of Lisieux to make the map, assisted administrators to gather local information, which may not have only highlighted their spiritual leadership to Rome, but also could be used to provide information for the centralized state and as a means of demanding attention and aid from that same centralized state. James B. Collins reminds the reader that "[l]ocal problems, like poor relief, came increasingly to be defined as national problems . . . *local elites* demanded help from the national government to meet relief obligations. The impetus for state intervention in poor relief did not come solely from the top down, it came from the bottom up."[91] As an instrument of polity then, d'Anville's map and its accompanying mémoire for this local diocese may have been part of the impetus from the "bottom up," but the bottom was still driven by the personal effort of an elite with little (or no) consideration for a commercial product, that is, subscription offers or broadsheet appeals. Although the map did not carry statistics on the number of poor in the diocese of Lisieux, it could provide a visual image for an accompanying mémoire with the necessary data provided by local parish priests; it was an instrument of the state at the local level. It was to these local amateurs that d'Anville turned for the geographical information for this map. D'Anville's *Mémoire Instructif* provided instructions for the bishop and his priests on the type of information to gather such as the number of "faithful," religious houses and affiliation, boundaries of villages and hamlets, details of the river system, forests, marshes, and vineyards, as well as an explanation of the framework for the maps to be made in the field.

Following Chevalier's method and the framework described by d'Anville in his *Mémoire instructif*, each local priest compiled a map of his parish and the surrounding area with the parish church at the center of the map and the furthest point two leagues out. Each local priest had his template on which to depict his surrounding villages, monasteries, hospitals, other municipal buildings, such topographical features as rivers, forests, and vineyards, and both ecclesiastical and civil boundaries. In addition, the priest was to mark the angle of the rising and setting sun from the equinoctial point and place this daily value along the eastern and western arc of the circular framework. Each map was to be accompanied by a mémoire that would give added historical information about the local place names, the parish, and the surrounding land. Through a process of revision as part of the "reasoned approach" of the Enlightenment, d'Anville again solicited the aid of the various parish priests to critique his compiled map and published a second *Mémoire Instructif* to assist his amateurs in the revision process. D'Anville's first initiative succeeded and, while it addressed a relatively small area, it helped to confirm the optimism of an Enlightenment ideology that the popularization of science, reasoned argument, and encyclopedism could indeed advance human knowledge and address social concerns.[92] Here was science in the service of the state and perhaps not at the mercy of the commercial trade.

Two years after his successful undertaking to map Lisieux, d'Anville reprised the process for the bishop of Blois, Monseigneur Caumartin. Since Blois was located in the ecclesiastical province of Paris, Jean François Paul Le Fèvre de Caumartin, from a prestigious robe family, may have solicited a map from d'Anville for the same reason that Brancas did (for Lisieux) in the 1720s. Both bishops, however, may also have sought a map for their ecclesiastical domain for the purposes of calculating the wealth of various parishes from which the bishop elicited funds for the *capitation* exemption. The *capitation*—a direct tax introduced in 1695—continued to be collected throughout the eighteenth century. The clergy, who were not intended to be exempt from this universal tax, negotiated a release or rather purchased an exemption for a set annual fee. The local parishes bore the burden of providing this fee. D'Anville's map and the reports from local parish priests could help the bishop determine the amount from each parish and the size of his contribution to the clergy exemption.

Following the same process for Caumartin, d'Anville drew base maps and wrote instructions to be distributed to all the parishes of Blois.[93] At the beginning of his new *Mémoire Instructif*, d'Anville assured the bishop of the benefit of such a method to his ecclesiastical administration and of the ease of usage for the most simple and least scholarly of his flock. The same circular template was provided for each local parish priest and, in turn, each parish provided d'Anville with an on-site map and mémoire. Although several manuscript maps of the various parishes of Blois are found in d'Anville's papers, the map of the diocese of Blois was never completed.[94] The patron of the map project, the bishop of Blois, died in 1733, and the ecclesiastical administration of the area remained unsettled for two years.

Despite the failure to complete the map for the diocese of Blois, d'Anville still advocated the method of using locals to make maps on-site, which he or any other *géographe de cabinet* (geographer of the study) could compile. Over twenty years after he first proposed the project to the bishop of Lisieux and just as the first national map survey was nearing completion (1730–1744), d'Anville published a more general version of his *mémoire* for drawing maps to serve a diocese or a province.[95] Two years later in 1745, the local intendant of Soissons, Charles-Blaise Méliand, ordered a reprinting of this work in order to have a map drawn of the *généralité* of Soissons. While the initiative appears to have stopped there, d'Anville's method could have been very helpful for Méliand facing governmental reformist efforts to improve the roads throughout France as well as Méliand's future reform efforts in manufacturing and agriculture. And yet the optimism of the reasoned approach and the Enlightenment ideology to advance human knowledge when faced with the economic realities stalled on the drawing board if it even reached the drawing board. Perhaps the convergence of two projects, one just completed and the other still under way, led Méliand to utilize the engineering work already accomplished. After all Cassini's survey of France was completed in 1744, and the survey sheets were printed between 1738 and 1744. In the decade prior Philibert Orry, the former intendant of Soissons (1722–1727) and current Director of the Ponts et Chausées [Bridges and Roads], had not only sent Jacques Cassini back into the field to complete the survey, but also set engineers (beginning in 1738) to work on producing maps of the road system of France.[96]

By the mid-eighteenth century after presenting the results (in 1744) from the first national mapping survey, César-François Cassini de Thury (Cassini III) was heading up a new national survey to map France at even greater detail (beginning in 1747) than that begun by his father, Jacques. Cassini de Thury had also published a work addressing mapping procedures so that a more detailed survey might be conducted at the regional level and then brought together at the national level. Using the 1745 resulting map sheets, he believed that local efforts could create a national masterpiece, but in 1747, Louis XV with greater military concerns asked Cassini to undertake the second survey project.

The local project in Languedoc had continued to falter. Philippe Buache (1700–1773), heir to Delisle's business, had taken over as the Parisian geographer to create the maps based on the drawings from the field, but there were tensions as engineers and members of the Society balked at Buache's intention to indicate in the title of the maps that they had been "subjected to trigonometric and geometrical observations" by Buache. These men argued that they had already conducted these observations quite ably in fact. Christian Licoppe suggests that "this was a clear mid-century instance of the authorship debates allowed by the ethos of astronomical and geometrical cartography . . . in which observers in the field could argue against the geographers for the right to authorship pertaining to maps."[97] In 1749, the president of the Estates even refused to turn the project over to the Cassini group beginning the second national mapping project. Buache continued to try to direct from Paris even sending a questionnaire to local clergy to assist him with information as he revised the maps for the Languedoc project. It was important for Buache to obtain information to fill in the area of the triangles as the triangulation network alone did not make a map, but created a geometric framework that needed to be filled in.[98] By the 1760s Buache had made little progress in completing the maps, and there remained concerns that what was completed did not meet the standards of the on-going national project which soon took it over.[99] Cassini de Thury had criticized the surveying and observational practices of the Montpellier academicians and declared that many of their triangles were off by one to ten degrees.[100] From Cassini de Thury's perspective it took the national project to inform the local project of its geographical identity. As Charles W. J. Withers writes, "mapping in the Enlightenment illustrates how maps constitute geographical identity rather than reflecting it."[101] By the last decades of the eighteenth century, the "geographical identity" of Languedoc was constituted in dialogue solely between the Estates and the Cassini engineers without the voice of the Montpellier academicians.[102]

Among the more successful dialogues was the one that took place between the Estates of Burgundy and Joseph-Dominique Seguin (d. 1791), a royal engineer-geographer working with Cassini de Thury. Just over forty years after the Estates negotiated with Delisle, Seguin was approached by the Estates General of Burgundy which contracted with him (5 August) in 1752 to produce a map of the province.[103] In a mémoire from 1762, Seguin described the process of making the map that included using information from the local clergy drawn from their completed questionnaires. The large map (in 15 sheets), *Carte Particulière du Duché de Bourgogne levée géométriquement par ordre de MM. les Elus généraux de la Province . . .*, was sold assembled or in separate sheets for 48 livres.[104] Élie Catherine Fréron advertised this map in his *L'Année Littéraire*, in which he praised the detail of the map and the quality of its engraving and reported that this beautiful map was available at Seguin's shop in Paris.[105] Drawn from the input of the members of the Estates, this highly praised large scale map—based on triangulation—carried a plan of the city of Dijon and the royal fleur-de-lis in an ethereal cloud at the top center of the map, with the arms of the Condé family in the lower left corner and an explanatory cartouche of symbols and the arms of the Estates in the lower right corner atop the title cartouche.[106] This work was part of a grander informational project evidenced by the Estates' plan to assess the roads in a map (also contracted with Seguin) which grew out of Orry's instructions in 1738, and the continuation of those efforts by Daniel Trudaine in 1744. In the midst of the second national mapping project, some provinces contracted with Cassini's engineers to produce their own regional maps.[107] This map of Burgundy was the most successful.

While Cassini de Thury was a bit too optimistic regarding the number of maps (ten) that could be completed each year, he understood that the key to success for the second survey project would be organization, and the legacy might very well be "the potential for making mapping a permanent and continual activity of government."[108] By the start of the Seven Years' War only two maps were complete (Paris and Beauvais) and once again war put a stop to funding as the controller general, Jean Moreau de Séchelles, argued that money was needed elsewhere, but granted the company formed by Cassini de Thury, a subvention of 150,000 livres. Louis XV gifted Cassini de Thury the instruments, maps, and documents as well as the engineers conducting the observations and measurements in the field and a thirty-year privilege to print the subsequent maps.[109] Cassini de Thury kept the project going by establishing a company and selling shares which attracted a great many elite patrons. While it was no easy feat and the engraving was still on going on the eve of the French Revolution, the state attention and provincial contracts were in great contrast to the un-systematic private enterprises of the county mapping projects in England. As Mary Pedley reports: "No county governments existed with fiscal powers to authorize and sustain such surveys. And while the benefits of large-scale mapping might have been contemplated by individual government ministers, the constantly shifting locus of decision making and fiscal authority in eighteenth-century England meant that support from the Crown and/or Parliament was erratic and minimal."[110]

COMMERCE AND SOCIAL PROMINENCE ON ENGLISH COUNTY MAPS

In England similar developments in public work projects prompted the development of new county maps, but rather than a center-directed project, local amateurs (few of whom were professional surveyors or part of an engineering corps) initiated these plans.[111] This effort was given greater impetus by the mid-eighteenth century, aptly named organization, Society for the Encouragement of Arts, Manufactures and Commerce, which offered a prize of 100 pounds for an accurate county survey at a scale of one inch to a mile.[112] Some national pride was at stake given that France had already completed one national survey (by 1744) and was embarking on a more detailed project in 1747, but it was clear that the development of the turnpike trusts (beginning late seventeenth century) and canal development prompted and required greater local geographical awareness throughout England.[113] The Society was interested in promoting the technical arts and upholding the utilitarian nature of science. At its first meeting in 1754 at the Rawthmill Coffee House in Covent Gardens, the members expressed concern for reliable surveys to serve the interests of commerce.[114] In fact, in September 1755 William Borlase (1695–1772), an antiquarian, wrote to his friend Henry Baker (1698–1774), a fellow antiquarian and, like Borlase, a fellow of the Royal Society, of his concerns for the "state of British geography."[115] Borlase—writing just two years after John Green expressed similar concerns—suggested that there was no "public encouragement" for the field of geography as there was "in some foreign Countries."

> Our maps of England and its counties are extremely defective . . . if among your premiums for Drawings some reward were offered for the best Plan, Measurement and actual Survey of City or District, it might move the Attention of the Public towards Geography, and in time, perhaps, incline the Administration to take this matter into their Hands (as I am informed it does in some foreign Countries) and employ proper Persons every year, from Actual Surveys, to make accurate Maps of Districts, till the whole Island is regularly surveyed . . . and attempts what would be so much for the Honour as well as Commerce of this Island.[116]

Nearly forty years before George III established the Ordnance Survey (in 1791), the "Honour as well as Commerce" of the island was in the hands of private individuals.[117]

Benjamin Donn, a local mathematician, was the first to register with the Society for the prize and began a survey of the county of Devon, a county map project Gascoyne was unable to accomplish in the opening years of the century due to lack of encouragement. Over the course of five and half years he carried out a survey of an area about 1/5 the size (6700 sq. km. or 2600 sq.

mi.) of the province of Burgundy.[118] He produced a map in 12 sheets about 6 x 6 feet in size and sold at several shops in London and by booksellers in Devon. While the plan to survey Devon was not part of a larger national project, it was part of a movement initiated by the London-based Society and engraved and published by London-based establishments. The goal was to produce county maps at a sufficient scale for detail and the possibility of bringing the maps together as a uniform whole. Rather than sending proof sheets to the local parishes, Donn sent his proof sheets to the prominent gentlemen of the county; "Gentlemen well acquainted with each respective Part," who could indicate what improvements were necessary.[119] The quality of the work was, he suggested, indicated by the use of latitude and longitude observations, the measurements of the angles and positions of prominent locations, 6000 miles of roads and rivers, and at least ten stations per mile to achieve the greatest accuracy in the trigonometric network.

Based on a trigonometric survey with a scale of one inch to a mile (1: 633360) and engraved by Thomas Jefferys, the map was well received and earned Donn the 100-pound prize for one of the first county maps.[120] The subscription list is informative as it indicates that nearly 75 percent of the subscribers were from the landed elite and 90 percent from the county. Similar to Seguin's map of Burgundy, Donn included the coats of arms of two prominent local elites, John Baring and Matthew Lee though interestingly not Lord Courtenay, the Earl of Devon who ordered only two copies compared to the one hundred copies each ordered by Baring and Lee.[121] Donn's map was comparable in cost to Seguin's at one and ½ guineas (or 1 pound 11 shilling and 6 pence) and if bound in a book for 2 guineas.[122]

Accompanying the map was a list of "subscribers, gentlemen and ladies, with the names of their seats or places of residence." Engraved by Jefferys, the map displayed a scenic title cartouche with a clerk sitting upon a package gesturing toward the title as a worker moving a cask looked on. A worker picked fruit from a tree forming the right side of the cartouche as a gentleman sitting upon a rock looked on and out over a harbor. The scene was reminiscent of the active commerce of this county. On the map, Donn assured that while the latitude and longitude values differed from earlier charts and those of the nearby county, his values were accurate: "The Author thinks himself obliged to assure the Publick that these are taken with such Instruments and Care, as to leave no doubt of their Accuracy." The list of symbols included those for a farm or cottage, a village, copper and tin mines, parks, Roman or Danish forts or encampments, Turnpike Roads or those slated to become such, open roads through commons or downs, principal seats, churches, market towns and capitals. The map carried insets for a plan of Plymouth Dock and the town of Stoke as well as the town of Plymouth and its Citadel and an inset of the Island of Lundy off the northern coast in the Bristol Channel, and the city of Exeter and its suburbs.

This map not only plays a role in the chapter of history on the Society of the Arts, but also in the commercial realities of the production of such works. Baring and Lee are noted for their encouragement in Donn's dedicatory cartouche, but the expense of producing county maps was potentially business-ending as was nearly the case for Thomas Jefferys (1719–1771) who engraved Donn's map and sold it at his shop in London.[123] Jefferys, who had engraved maps for the *Gentleman's Magazine*, had already been involved in publishing the *Small English Atlas* (1749), "a correct sett of portable pocket Maps of all the Counties in England and Wales." The "encouragement" given for this work was according to the publishers, Jefferys and Thomas Kitchin, "the strongest proof we can desire that it is the neatest the correctest [sic] and cheapest of the kind ever yet publish'd."[124] The frontispiece carried no armorial bearings but depicted Mercury delivering an atlas to Britannia while Abundance looked on. With the commercial market still served with such editions obtaining enough subscribers would be challenging. With the end of the Seven Years' War and Jefferys' prior connections to the government for maps of America (see chapter 4), he may have heard of General William Roy's first formulations for a national survey.[125] Moreover, he was aware of the Society's incentive of 100 pounds for accurate county surveys. Jefferys, who had been so successful with his American maps, soon involved himself with more domestic products. Unlike other London map makers who were pleased to simply sell the county map, have it engraved, or buy the plates after the first edition of the map, Jefferys directly involved himself in the process from the start.[126]

He invested in several surveys directly and acted as engraver for others totaling as many as ten different county map projects.[127] Jefferys had been involved in directing the survey of Bedfordshire in 1765 and perhaps, even, "delayed" the completion of the engraving of Donn's map of Devon, which was ultimately the first to win the Society's award. Having so recently benefited from his government connections in producing his American maps, Jefferys soon found himself financially overburdened as he took on the various stages of the county map project which entailed much expense prior to the process of engraving, Jefferys' forte.[128] One year after the publication of Donn's map of Devon, Jefferys declared bankruptcy.[129] These were challenging projects for individuals to carry out.

Nearly one hundred years after Ogilby's map of Essex was published, Thomas Sparrow initiated a plan to survey the county anew. Two years later, in 1772, John Chapman, a surveyor from county Suffolk, and Peter André took over the project left idle by Sparrow. In 1777, they published a work considered one of the most accurate of its genre before 1780.[130] Harkening back to the work of Gascoyne at the end of the previous century, Chapman's map of Essex likely emerged partly out of his estate map project for the rector of the parish of Beaumont-cum-Moze. He had also partnered with the widow of John Rocque to produce new maps and had been involved in the creation of the illustrations of country seats in Essex for the work, *History of Essex*. The surveying took three years, from 1772 to 1774, but the publication did not occur until 1777 leaving some of the proprietors of the principal seats inaccurate or out-of-date. As a surveyor, Chapman had been quite busy with proposing, surveying, engraving, publishing, or revising other county maps such as Yates' Lancashire and Staffordshire, Burdett's Derbyshire (originally from 1767), and even his own map of Lincolnshire. The map of Essex in 26 sheets (22 just for the map portion) at a scale of 2.5 inches to the mile was a spectacular work carrying not only a separate sheet for the list of 216 subscribers of which 75 percent were of the landed elite, but also a plan of the city of Colchester, and a view of the harbor of Harwich.[131] In late 1778, when Chapman died, he left several projects unfinished including the full print run of the map of Essex.[132] Chapman had been more successful than most in accomplishing his plan which was likely due to the fact that he was a surveyor having his hands in more than one project.[133]

Though not a surveyor, in the late 1770s, John Prior (1729–1803) met with success with his survey of Leicestershire and was awarded by the Society a silver medal and twenty guineas.[134] Prior, curate mathematician and school master, had enjoyed the patronage of the Hastings family and by their "encouragement" had obtained his degree in divinity in 1772.[135] Unlike Chapman, Prior's amateur status had led to time limitations for his surveying, and he was forced to rely on an earlier map of 1722.[136] Nonetheless he carried out a triangulation survey of the area. The map adorned with Francis Hastings' armorial bearing as the Earl of Huntingdon (1729–1789), a fellow of the Royal Society and a member of the Society of Antiquarians, was published in London and in Leicestershire by Prior. Prior included on the map a diagram of his triangulation stations, "[i]ntended for the use of the curious, and for a testimony of the accuracy of this work." The symbols on the map include those for market towns, villages, seats, farmhouses, churches and chapels, wind and water mills, coal pits, lime works, turnpike roads, cross roads, boundaries of the hundreds, and roman stations. There are small stands of trees (i.e., woods) and large forests depicted, and parks. In the lower left corner is a plan of Leicester with prominent buildings noted such as the castle, infirmary, churches, and the exchange. Similar to Chapman's map, the list of 264 subscribers indicated that the majority were from the county and made up of nobles, gentry, clergy and members of the growing professional classes.[137] Despite Prior's success it remained a challenge for these individual initiatives to succeed without an infrastructure to oversee the quality, uniformity, and finances that such endeavors required.

CONCLUSION

With the War of American Independence ended, in 1783 Cassini de Thury proposed to the British government to extend the geodetic program of France to its shores, but his death the following year seemed to delay any joint activities until 1787. The fate of the French project as the Revolution began was uncertain though engraving and editing did continue. In 1793 the Convention transferred the map to the army reportedly to ensure "its survival and its utility for the state and preventing domestic or foreign foes from seizing it."[138] As the map was a royal project to serve the needs of the state, the new government initially had no intentions of compensating the company created by Cassini de Thury over thirty years earlier. Cassini IV caught between the operations of the Old Regime and a new modern bureaucratic state (no doubt, whose foundations had been set earlier) resigned from the Observatory.[139] While Napoleon would later order that shareholders of the company be compensated, it was of little consolation to Cassini IV who as Konvitz suggests failed to recognize that the work of his family had succeeded in making mapping a routine exercise of government.[140] With the professionalization of cartography, when the government intended to begin a new surveying effort in the second decade of the nineteenth century, it looked no further than its own bureaus of professional cartographers to serve the utility and security of the *modern* state.

After the death of Cassini de Thury, his son Jean Dominique Cassini (Cassini IV) would communicate with members of London's Royal Society. Joseph Banks, president of the Royal Society, William Roy and other fellows began the geodetic program for England by measuring the baseline at Hounslow Heath (1784) for the triangulation network. Roy re-initiated his plans for a complete survey of the British Islands and reported to the Royal Society the importance of such work: "Accurate surveys of a country are universally admitted to be works of great public utility, as affording the surest foundation for almost every kind of internal improvement in time and peace, and the best means of forming judicious plans of defence [sic] against invasions of an enemy in time of war."[141]

He had had at least two decades to contemplate the operations and had even written to the Duke of Richmond, head of the Board of Ordnance, detailing the project.[142] Richmond had already employed two surveyors, Thomas Yeakell and William Gardner in a county map project, the *Great Survey* of his own Sussex County which grew out of their earlier work on estate mapping. Only four sheets were completed by the early 1780s, as Yeakell and Gardner had been taken in to the surveying of the Board of Ordnance in the beginning stages of the Ordnance Survey (1791). After Yeakell's death in 1787 and Roy's death in 1790, Gardner who was soon attached to the Ordnance Survey returned to complete the county map of

Sussex. In a similar situation to Seguin, Gardner, a professional surveyor attached to the national project completed the Sussex map and dedicated it to Richmond in 1795. It was among the last of its kind as the Ordnance Survey replaced individual initiatives.

The surveyors of the Ordnance Survey would return to the works produced in the last forty years and recognize the efforts to well utilize instrumentation to build a triangulation network that could then be extended to the other areas of the British Islands yet to be covered by this trigonometric network. In answer to Borlase's and Green's concern that England was lagging behind, a central office was now in place to build upon the works of private initiative that had laid the groundwork. While earlier small scale county maps and the atlases compiled from them that had saturated the market became quaint ephemera in the nineteenth century, surveyors moved to the invasion coasts and the triangulation of south-east England to begin the network of the Ordnance Survey. Just as England and France had done for centuries, they moved to protect their greatest asset, the English Channel and the topic of the next chapter.

NOTES

1. Marie Noëlle Bourguet, Christian Licoppe, and H. Otto Sibum, eds., *Instruments, Travel and Science: Itineraries of Precision from the Seventeenth to the Twentieth Century* (New York: Routledge, 2002), 8–9; Josef Konvitz, *Cartography in France, 1660–1848: Science, Engineering, and Statecraft* (Chicago: University of Chicago Press, 1987), 2; and also see Mary Sponberg Pedley, *The Commerce of Cartography: Making and Marketing Maps in Eighteenth-Century France and England* (Chicago: University of Chicago Press, 2005), 25–34.

2. James E. McClellan III and François Regourd, *The Colonial Machine: French Science and Overseas Expansion in the Old Regime* (Turnhout: Brepols, 2011), 72–73.

3. This title was created (in 1718) for Delisle to set him apart from the title of *géographe du roi* (royal geographer) which carried a smaller pension but was accorded to engravers and map publishers (e.g., Jaillot and Nolin). This "premier" title was to be a mark of distinction and was held by Delisle until his death at which time it was accorded to d'Anville.

4. Pedley, *Commerce of Cartography*, 33.

5. John Brian Harley, "Power and Legitimation in the English Geographical Atlases of the Eighteenth Century," in *Editing Early and Historical Atlases*, ed. Joan Winearls (Toronto: University of Toronto Press, 1995), 173.

6. Konvitz, *Cartography in France*, 4, 9, 22.

7. Pedley, *Commerce of Cartography*, 33.

8. John Green, *Remarks in support of the new chart of North and South America* (London, 1753), 3.

9. Pedley, *Commerce of Cartography*, 34.

10. P. D. A. Harvey and H. Thorpe, *The Printed Maps of Warwickshire 1576–1900* (Birmingham: University of Birmingham, 1959), 19; and Andrew Macnair and Tom Williamson, *William Faden and Norfolk's 18th-Century Landscape* (Oxford: Oxbow Books, 2010), 6.

11. Ida Darlington and James Howgego, *Printed Maps of London, circa 1553–1850* (London: George Philip & Son Limited, 1964), 22.

12. John Ogilby, *Britannia*, facsimile edition (Amsterdam: Theatrum Orbis Terrarum LTD, 1970), xii, xiv.

13. John Ogilby, *A New Map of Kent* (London, 1672), BL, Maps *3055 (37).

14. John Ogilby, *Midlesex* (London, 1677), BL, Maps *3455 (7).

15. John Ogilby, *Essex actually surveyed* (London, 1678), BL, Maps *2345(1). Gregory King drew this map from survey, but Morgan published it, and it carried Ogilby's name. Ogilby, *Britannia*, facsimile, xii; and Catherine Delano-Smith, "Milieus of Mobility: Itineraries, Route Maps, and Road Maps," in *Cartographies of Travel and Navigation* James R. Akerman, ed. (Chicago: University of Chicago Press, 2010), 67.

16. Linda Levy Peck, *Consuming Splendor: Society and Culture in Seventeenth-Century England* (Cambridge: Cambridge University Press, 2005), 272–73.

17. See Andrew McRae, "To Know One's Own: Estate Surveying and the Representation of the Land in Early Modern England," *Huntington Library Quarterly* 56, no. 4 (1993): 333–57.

18. Some of the material for regional mapping in France is drawn from my first book, *When France was King of Cartography: The Patronage and Production of Maps in Early Modern France*, 36–40. Also see François de Dainville, *Cartes Anciennes de l'Eglise de France* (Paris: Librairie Philosophique J. Vrin, 1956), 11.

19. Jean Jubrien, *Carte du pays et diocese de Reims* (Chaalonnois, 1623), BN, Cartes, GE DD 2987 (292B).

20. Dainville, *Eglise*, 23–24; also see the dedication on the map, Jubrien, BN, Cartes, GE DD 2987 (292B). "Ces descriptions ne sont pas seulement utilles aux capitaines qui manient les armes matérielles mais aussy aux prélats de l'Eglise qui sont préposez sur les spirituelles, car comme ainsi soit que l'Eglise est semblable à une armée toujours prête à combattre, . . . pour recognoistre si tout y est en point de combattre contre le diable, le monde, et la chair, nos communs ennemis ce qu'ils ne peuvent faire plus commodément qu'à l'ayde d'une carte de leur diocese, sur laquelle ils puissant voir d'un seul traict d'oeil la situation."

21. Dainville, *Eglise*, 31.

22. See Mireille Pastoureau, *Nicolas Sanson d'Abbeville: Atlas du Monde 1665* (Paris: Sand & Conti, 1988).

23. Dainville, *Eglise*, 41.

24. Jean Baptiste Colbert, "Instructions pour les maîtres des requêtes, commissaires départis dans les provinces, Septembre 1663," in *Lettres, instructions et mémoires de Colbert*, ed. Pierre Clement (Paris: Imprimerie Impériale, 1867), 4: 27–43.

25. Konvitz, *Cartography in France*, 2.

26. Derek Howse, *Greenwich Time and the Discovery of the Longitude* (Oxford: Oxford University Press, 1980), 16.

27. A. M. de Boislisle, ed., *Mémoires des Intendants sur L'État des Généralités dresses pour l'instruction du duc de Bourgogne*, Book One, *Mémoire de la Généralité de Paris* (Paris: Imprimerie Nationale, 1881). http://gallica.bnf.fr/ark:/12148/bpt6k64389722/f484.image.

28. Kenneth J. Banks, *Chasing Empire Across the Sea: Communications and the State in the French Atlantic, 1713–1763* (Montreal and Kingston: McGill-Queen's University Press, 2002), 88. Quoting from Daniel Nordman and Jacques Revel, Banks writes that "knowledge of territory became inseparable from the exercise of sovereignty." I would suggest this sentiment took hold with the Colbertian programs of the second half of the seventeenth century.

29. Dainville, *Eglise*, 56–57. On intendant Basville, see W. Gregory Monahan, "Tyrant of Languedoc? Nicolas de Lamoignon de Basville in Public and in Private," *Proceedings of the Western Society for French History* 37 (2009): 23.

30. Hubert Gautier, *Carte du Diocése de Thoulouse* 1695. BN, Cartes, GE DD 2987 (356) drawn map with dedication to Colbert by Gautier.

31. Alexis Hubert Jaillot, *Le Diocese de Toulouse* (Paris, 1695), BN, Cartes, GE DD 2987 357B.

32. Jean Baptiste Nolin, *Le Diocese de l'Eveché de Nismes Dressé Nouvellement sur les Lieux par le Sr. Gautier Architecte et Ingenieur de la Province de Languedoc Dedié a Monseigneur Esprit Flechier Evêque de Nismes par son tres humble et tres obeisant serviteur* (Paris, 1698), BN, Cartes, GE DD 2987 (354).

33. Monahan, "Tyrant of Languedoc," 21–34.

34. In the small cartouche with the explanatory symbols, Nolin included his shop address as well as this statement: "Cette carte a esté faite par l'ordre et par les soins de Mgr. Flechier eveque de Nismes l'an 1698."

35. A decorative banner explained that Nismes, once a Roman colony, was the oldest city of Gaul. In addition perhaps as a means of explaining the importance of the map, the area had been recently (16 May 1694) divided to create the diocese of Alais, the fourth ranking city of the Estates of Languedoc. The map also reveals the borders from the Mediterranean to the south, the marshlands to the southeast beyond the Rhone River, Provence to the east, the diocese of d'Uzes to the north and the "new" diocese of D'Alais to the northwest and the Diocese of Montpellier to the southwest.

36. See William Beik, *Absolutism and Society in Seventeenth-Century France: State Power and Provincial Aristocracy in Languedoc* (Cambridge: Cambridge University Press, 1985), 115–16. Beik writes about the administrative realities for intendants in the region of Languedoc. "Languedocian intendants [agents of the crown], therefore, had to work especially closely with governors and other negotiators [ecclesiastical elites], and their success was especially embroiled in political factionalism because the Estates had to be won over by negotiation and influence."

37. Jean Baptiste Nolin, *Le Gouvernement Général de Languedoc . . .* (Paris, 1697), BN, Cartes, GE DD 2987 (650). "En un mot cette Carte est en beaucoup d'endroits plus exacte que celles qui ont paru jusqu'a present."

38. Petto, *When France was King*, 155–56.

39. Lloyd A. Brown, *The Story of Maps* (Boston: Little, Brown Company, 1950), 246.

40. Messieurs de l'Académie, *Carte particulière des environs de Paris* (Paris, 1678), BN, Cartes, GE DD 2987 (788, I–IX).

41. Albert Jouvin de Rochefort, *Diocèse de l'Archevesché de Paris* (Paris, 1675), BN, Cartes, GE D 16898. Also see Dainville, Eglise, 52.

42. Alexis Hubert Jaillot, *La Généralité de Paris divisée en ses esclections* (Paris, 1708), BN, Carte, GE 2987 709B.

43. Jean Phélypeaux, *Mémoire de la Généralité de Paris* (Paris: Imprimerie Nationale, 1881), xi–x. http://gallica.bnf.fr/ark:/12148/bpt6k64389722/f484.image.

44. Alexis Hubert Jaillot, *Evesché de Meaux dedié a Monseigneur Jacques Benigne Bossuet, evesque de Meaux par son tres-humble et tres-obeissant serviteur Hubert Jaillot, g.o.d.r.* (Paris, 1698), BN, Cartes, GE DD 2987 (279B). See Dainville, *Eglise*, 54–55. Dainville writes that the 1698 print is rare and most are dated 1701 or 1717.

45. Jean Besson, *L'Archeveche de Paris divise en ses trois Archidiaconez et en ses deux Archiptretrez et sept doyennez ruraux. Dresse et mis au jour par ordre de son eminence Monseigneur Le Cardinal de Noailles* (Paris, 1706), BN, Cartes, GE DD 2987 (277, I-IV). The map carried the coat of arms of Noailles and ecclesiastical symbols. See Dainville, *Eglise*, 55.

46. See Nelson-Martin Dawson, *L'Atelier Delisle: L'Amérique du Nord sur la table à dessin* (Sillery, Québec: Editions du Septentrion, 2000).

47. Guillaume Delisle, *Carte du Diocese de Beziers dressée sur les lieux par le S Gautier Ingenieur et Architecte de la Province des Languedoc, et rectifiée sur les observations de l'Academie Rle. Des Sceinces par G. Delisle de la même Academie a Monseigneur Monseigr Charles des Alris de Rousset evêque et Seigr de Beziers* (Paris, 1708), BN, Cartes, GE DD 2987 (350).

48. In 1704, two royal decrees were issued requiring that no map could be printed in the kingdom without the publisher/editor having acquired permission or the privilege of the royal seal and eight copies of the work distributed to the required royal officials including the chancellor and royal librarian. In addition, no geographer could take the *quality* of royal geographer without the official brevet. These were prompted by plagiarism suits including the one between Nolin and Delisle. In his legal dispute with Nolin, Delisle expressed concern that the important science of geography not fall under the control of copperplate engravers. See Petto, *When France was King*, 156–57.

49. Guillaume Delisle, *Carte topographique du diocese de Senlis levée sur les lieux par Mr. Parent cure d'Aumont* (Paris, 1709), BN, Cartes, GE DD 2987 (303). See Dainville, *Eglise*, 58.

50. Petto, *When France was King*, 62–63.

51. See AN, Marine, 6JJ 71, *Manuscrits de Delisle*, for manuscript maps used to produce the map of Burgundy. Also see François de Dainville, "Cartes de Bourgogne du XVIIIe siècle,"

Actes du 84e Congrès National des Sociétes Savantes, Dijon, 1959 (Paris: Comité des travaux historiques et scientifiques, 1960), 1–7.

52. Guillaume Delisle, *Carte du Duché de Bourgogne et des Comtez en dependans . . .* (Paris, 1709), BN, Cartes, GE DD 2987 (502–3).

53. AR, Dossier, "Guillaume Delisle."

54. Christian Licoppe, "The Project for a Map of Languedoc in Eighteenth-Century France . . . ," in *Instruments, Travel and Science: Itineraries of Precision from the Seventeenth to the Twentieth Century*, ed. Marie Noëlle Bourguet, Christian Licoppe, and H. Otto Sibum (New York: Routledge, 2002), 53.

55. For this section on Languedoc, see François de Dainville, *Cartes Ancienne du Languedoc XVIe-XVIIIe siècles* (Montpellier: Societé languedocienne de géographie, 1961), 90–110. Also see Licoppe, "Map of Languedoc," 51–74.

56. Licoppe, "Map of Languedoc," 54.

57. Licoppe, "Map of Languedoc," 54.

58. For the negotiations between Montferrier (the syndic of the province representing the Estates) and Delisle's widow, see Dainville, *Languedoc*, 108; Licoppe, "Map of Languedoc," 55–56; and Petto, *When France was King*, 66–67.

59. Licoppe, "Map of Languedoc," 56–57.

60. See W. Ravenhill, "Joel Gascoyne, a Pioneer of Large-Scale County Mapping," *Imago Mundi* 26 (1972): 60–70. Also see Macnair and Williamson, *William Faden*, 6.

61. Ravenhill, "Joel Gascoyne," 65.

62. MacNair and Williamson, *William Faden*, 1.

63. Ravenhill, "Joel Gascoyne," 61. At the death of his patron, Gascoyne ceased his membership in the Drapers' Company and his mapping took a more terrestrial direction as he devoted his attention to land surveying from 1690 to 1703.

64. Ravenhill, "Joel Gascoyne," 66. His surveying efforts produced the *Lanhydrock Atlas*, which included surveys of 258 properties.

65. Ravenhill, "Joel Gascoyne," 65.

66. Ravenhill, "Joel Gascoyne," 65.

67. Ravenhill, "Joel Gascoyne," 65. He had included as many locations "without making it burthensom to the eye; all which are in their true places and distances; and I presume to say, with as few errors as in such undertakings."

68. Ravenhill, "Joel Gascoyne," 65.

69. Joel Gascoyne, *A Map of County Cornwall Newly Surveyed* (London, 1699), BL, Maps CC.5.a.279.

70. Ravenhill, "Joel Gascoyne," 67–69.

71. Harley, "Power and Legitimation," 169–71.

72. Harvey and Thorpe, *Maps of Warwickshire*, 19. Harvey and Thorpe write that Senex was a high caliber surveyor and map maker who was dissatisfied with the "standard of mapping" and undertook costly surveying efforts to produce large-scale works on the counties and to improve geographical knowledge. John Senex, *A Map of the County of Surrey* (London, 1729), BL, Maps K.Top.40.5. The coats of arms depicted on the map are for: Bishop of Canterbury; Bishop of Winchester; Lord King; Duke of Leeds; Duke of Newcastle; Duke of Argyle; Earl of Lincoln; Earl of Plymouth; Earl of Ailesford; Earl of Waterford; Viscount Windsor; Viscount Shannon; Viscount Palmerston; Lord Howard; Lord North & Grey; Lord Guilford; Lord St. John, Lord Monmouth; Lord Trevor; Lord Butler; Lord Onslon; Lord Baltimore; Sir Jacob Astley, Bart.; John Barber, Esq.; John Michel, Esq.; Anthony Tate, Esq.; Sam Horsey, Esq.; James Lambe, Esq.; Francis Child, Esq.; Humphrey Parsons, Esq.; William Feilding, Esq.; John Nicholas, Esq.; Zouch of Okin; John Weston, Esq.; John Temple, Esq.; Joseph Gascoigne, Esq.; John Heath, Gent.; Latton of Latton, St. George; Dr. Cranmer; Jason Oglethorpe, Esq; Thomas Steavens, Esq; Sir Mathew Decker, Bart.; Arthur Onslow, Esq.

73. Laurence, Worms, "The Search for John Senex F.R.S.: An Aspect of the Early Eighteenth-Century Book Trade," Paper delivered to the Bibliographical Society 18 January, 2000 (accessed 30 May 2014) (posted 30 January 2014) http://laurenceworms.wordpress.com/2014/01/30/the-search-for-john-senex-f-r-s-an-aspect-of-the-early-eighteenth-century-book-trade/.

74. Laurence Worms, "The maturing of British Commercial Cartography: William Faden (1749–1836) and the Map Trade," *The Cartographic Journal* 41, no. 1 (2004): 7. For the survey of Sussex County by Richard Budgen, Senex engraved the large county map (1724) dedicated to Spencer Compton, a knight of the shire. Similarly, he engraved parts of the map for the survey by William Williams of the two Welsh counties, Denbighshire and Flintshire. And he accepted subscriptions for Henry Beighton's map of Warwickshire.

75. See G. Walters, "Themes in the Large Scale Mapping of Wales in the 18th Century," *The Cartographic Journal* 5, no. 2 (1968): 136.

76. Walters, "Mapping of Wales," 135; and Worms, "Maturing of British Commercial Cartography," 7.

77. Worms, "Maturing of British Commercial Cartography," 7.

78. Harvey and Thorpe, *Maps of Warwickshire*, 19–20.

79. Shelley Costa, "Marketing Mathematics in Early Eighteenth-Century England: Henry Beighton, Certainty, and the Public Sphere," *History of Science* 40 (2002): 219.

80. Henry Beighton, *Proposals for Publishing by Subscription, a New, Large, and Correct Map of Warwickshire* (London, 1722).

81. Harvey and Thorpe, *Maps of Warwickshire*, 23.

82. Harvey and Thorpe, *Maps of Warwickshire*, 25.

83. Henry Beighton, *A Mapp of Warwickshire, . . . Actually Survey'd in the Years 1722, 1723, 1724, 1725* (London, 1725), BL, Maps, K. Top. 42.74.8.tab.end. See Harvey and Thorpe, *Maps of Warwickshire*, 25. They note that there were places for 152 shields and names along the left and right sides for a total of 304. On the map I examined at the British Library, there were 254 shields with names (and/or decorated arms). "The arms of the local gentry were added or erased in these spaces as they paid or defaulted in their subscriptions; accordingly the number of shields actually shown varies on different copies of the map." (25)

84. Herman Moll, *A New Description of England and Wales* (London, 1724), iii.

85. François Chevalier, "Sur un Manière de Lever la Carte d'un Pays," *Histoire de l'Académie royale des sciences* (1707): 141–45.

86. Konvitz, *Cartography in France*, 19.

87. Konvitz, *Cartography in France*, 19.

88. Jean Baptiste Bourguignon d'Anville, *Carte topographiques du Diocèse de Lizieux* (Paris, 1730), BN, Cartes, GE DD 2987 (256B). D'Anville also produced an accompanying text, *Mémoire instructif, pour faire la carte du diocese de Lisieux*. Also see Dainville, *Eglise*, 64–75.

89. Dainville, *Eglise*, 65-66.

90. James B. Collins, *The State in Early Modern France* (Cambridge: Cambridge University Press, 1995), 179, 184.

91. Collins, *State in Early Modern France*, 185.

92. Matthew Edney, "Reconsidering Enlightenment Geography and Map Making: Reconnaissance, Mapping, Archive," in *Geography and Enlightenment* ed. David N. Livingstone and Charles W. J. Withers (Chicago: University of Chicago Press, 1999), 186–87.

93. Jean Baptiste Bourguignon d'Anville, *Mémoire instructif, pour que, dans toutes les paroisses d'un diocese* (Paris, 1732). Also see Dainville, *Eglise*, 65–70.

94. Manuscript map of Blois found at BN, Cartes, GE DD 2987 (1245).

95. Jean Baptiste Bourguignon d'Anville, *Mémoire instructif, pour dressé sur les lieux des cartes particulières et topographiques d'un canton de pays, renferment dix ou douze paroisses* (Paris, 1743).

96. Stéphane J. L. Blond, "The Trudaine Atlas: Government Road Mapping in Eighteenth-Century France," *Imago Mundi* 65, no. 1 (2013): 64–79; and Konvitz, *Cartography in France*, 16.

97. Licoppe, "Map of Languedoc," 64.

98. Charles W. J. Withers, *Placing the Enlightenment: Thinking Geographically about the Age of Reason* (Chicago: University of Chicago Press, 2007), 102.

99. Petto, *When France was King*, 67; and Licoppe, "Map of Languedoc," 65.

100. Licoppe, "Map of Languedoc, 65.

101. Withers, *Thinking Geographically*, 102.

102. Licoppe, "Maps of Languedoc," 66.
103. Dainville, *Eglise*, 90.
104. Joseph-Dominique Seguin, *Carte Particulière du Duché de Bourgogne* (Paris, 1765), BN, Cartes, GE DD 2987 (504, I–IV B).
105. Elie Catherine Fréron, "Carte de Bourgogne," *L'Année Litteraire* 1(1765): 143–44.
106. Pedley, *Commerce of Cartography*, 59.
107. Pedley, *Commerce of Cartograhpy*, 81.
108. Konvitz, *Cartography in France*, 22.
109. Pedley, *Commerce of Cartography*, 80–81; and Konvitz *Cartography in France*, 24.
110. Pedley, *Commerce of Cartography*, 81.
111. Macnair and Williamson, *William Faden*, 19–20; and See J. B. Harley, "The Remapping of England 1750–1800," *Imago Mundi* 19 (1965): 56–67. The men who carried out the surveys not only often lived in the county they surveyed but also were not professional surveyors. As Harley notes the role of the London business establishments was in the "finishing process of county map-making." "Although the provincial surveyor had the skill and could acquire the capital to undertake field survey, very often there was no specialist map-engraver and printer in the provincial town. Because of this, or for reasons of prestige, the amateur turned to the professional" (65). Most of these new surveys were published and sold in London as well as in their respective counties.
112. Harley, "Remapping," 61. Between 1759 and 1801 only twenty-three county maps were submitted for the award.
113. Macnair and Williamson, *William Faden*, 14–15.
114. Macnair and Williamson, *William Faden*, 15. Also see J. B. Harley, "The Society of Arts and the surveys of the English counties 1759-1809," *Journal of the Royal Society of Arts* 112, no. 5090 (1964): 123. The concern was "for the Ease and Advancement of the National Commerce."
115. Macnair and Williamson, *William Faden*, 15.
116. Macnair and Williamson, *William Faden*, 15.
117. Mary R. Ravenhill and Margery M. Rowe, eds., *Maps of Georgian Devon* (Exeter: Short Run Press, 2002), 3–4. Ravenhill and Rowe suggest that after 1750 there was increase in map consciousness, "which reflected not only the landowner's desire to record details of his own estates, but his interest in what was happening elsewhere in the county by way of agricultural improvements, the landscaping of parks and the re-building of houses." (4)
118. Benjamin Donn, *Proposals for Publishing a New and Accurate Map of the County of Devon* (London, 1764). Donn reported in an address to the public, "This map of the County of Devon, now presented for your Acceptance, is made from an actual survey, on a Plan approved of by the Society *for Encouragement of Arts, Manufactures and Commerce*: And as no *Labour* or *Expence* has been spared, to render it as accurate and elegant as any Thing of the Kind can reasonably be expected, (having made it may entire Business from *August* 1759, to this Time) I flatter myself it will please the Candid." These "alterations" being completed he presented the map, January 10, 1765.
119. Donn, *Proposals for publishing a new and accurate map of the county of Devon.*
120. Benjamin Donn, *A Map of the County of Devon* (London, 1765), BL, Maps, *2119.5.
121. At the top of the map is a dedicatory cartouche with heraldic representations: "To John Baring of Mount Radford & to Matthew Lee of Ebford near Exeter, Esqrs. This Map, as a grateful Acknowledgement of the great and generous, assistance which they have given thereto, Is humbly Inscribed, by their most obedient servant Benjamin Donn." John Baring (1730–1816) was a successful merchant banker in Devon (bank was in Exeter) and future member of Parliament. Matthew Lee of Ebford was a local elite, his family having obtained their coat of arms in 1759.
122. Macnair and Williamson, *William Faden*, 40. If one pound is about 23 livres then the two maps are roughly comparable in price with 2 pounds about 46 livres. A rough approximation of the 500 or so subscribers indicates that 75 percent were of the elite landed class. According to Ravenhill and Rowe, there were 528 subscribers, of which 90 percent were from Devon.

123. See Laurence Worms, "Thomas Jefferys (1719-1771): Beginning the World Afresh," *Mapforum* 3 (2004): 20–29; and J. B. Harley, "The Bankruptcy of Thomas Jefferys: An Episode in the Economic History of Eighteenth Century Map-Making," *Imago Mundi* 20 (1966): 42–44.

124. Thomas Jefferys and Thomas Kitchin, *Small English Atlas* (London, 1749), preface.

125. William Roy, "An Account of the Measurement of a Base on Hounslow-Heath," *Philosophical Transactions of the Royal Society* 75 (1785): 385–480, in particular see 387–88. Also see Rachel Hewitt, *Map of a Nation: A Biography of the Ordnance Survey* (London: Granta Books, 2011).

126. Harley, "Remapping," 64.

127. Harley, "Remapping," 64; and Harley, "Bankruptcy," 42. Jefferys had invested in several surveys, that is, Bedfordshire, Buckinghamshire, Huntingdonshire, Oxfordshire, and Nottinghamshire, not all of which were successfully completed or produced maps.

128. Harley, "Bankruptcy," 42–44. As Harley mentions this was a costly endeavor as not only the task of hiring skilled surveyors, but getting out a proposal to obtain subscribers and booksellers that demanded a commissions to take in these subscriptions, instrument costs and other material prior to the actual engraving could require hundreds of pounds output before the map even arrived at the engraver's table. Harley provides the proposal sheet for the map of Huntingdon which was supposed to be 'out' in 1766 and was not published until four years later. (43)

129. Harley, "Bankruptcy," 28. As Harley notes: "At the very least Jefferys' misfortune provides a further illustration of the power of commercial considerations in determining the quality of maps: it was often the organization of the market, rather than technical feasibility, which inhibited cartographical progress." Harley, "Remapping," 63. These were expensive endeavors and subscriptions were often necessary to pull them off. Subscriptions lists indicate that 200 or more subscribers were achieved, and Harley notes that the first edition of these maps were about 300 copies. Donn's Devon project cost 2000 pounds so while the 100 pound prize was nice, it did not assuage the cost of the project.

130. John Chapman and Peter André, *A Map of the County Essex* (London, 1777), BL, Maps, 188.L.1. There are 25 sheets to this map and a 26th sheet with the smaller one sheet map of the county. Also see A. Mason, *Wasn't It exciting!: A Compilation of the Work of A. Stuart Mason* (London: Royal College of Physicians, 2004), 181–82.

131. About 60 percent of the subscribers were of the gentry class, about 15 percent of the noble class, 10 percent nongentry, 5 percent clergy, and 10 percent military officers. This roughly matches Donn's subscription list. See Mason, *Exciting*, 183.

132. Mason, *Exciting*, 186.

133. Mason, *Exciting*, 186.

134. John Prior, *Map of Leicestershire* (London, 1779), BL, Maps, *3260.(7.). The dedicatory cartouche reads: *To the Right Honourable Francis Earl of Huntingdon, Baron Hastings, Hungerford, Newmark, Peverel, Botreaux, Molins, Moels, and de Homet. By his Lordships permission this map of Leicestershire from an actual survey began in the year 1775 and finished in the year 1777 with the utmost gratitude and respect is dedicated by his lordships most obliged most obedient and most humble servant John Prior.*

135. See Basil L. Gimson and Percy Russell, *Leicestershire Maps, A Brief Survey* (Leicester: Edgar Backus, 1947), 19.

136. Gimson and Russell, *Leicestershire*, 19.

137. Macnair and Williamson, *William Faden*, 19.

138. Konvitiz, *Cartography in France*, 30.

139. Konvitz, *Cartography in France*, 29.

140. Konvitz, *Cartography in France*, 30-31.

141. Roy, "Hounslow Heath," 385.

142. See Pedley, *Commerce of Cartography*, 84.

Chapter Three

Chart Making in England and France, and Charting the English Channel

> [T]he faultiness of the Charts [is partly due to] the Compilers and Publishers, who publish to the world the rude Sketches of the Navigator as Accurate surveys, without telling what authority they have for so doing. . . . Neither can I clear Seamen of this fault. . . . I have known them [to] lay down the line of a Coast they have never seen, and put down Soundings where they never have sounded; and, after all, are so fond of their performances as to pass the whole off as Sterling. —*Captain Cook's Journal*, 7 September 1770

This chapter presents the history of the development of maritime atlases and charts in England and France from the sixteenth to the eighteenth century as well as the historical context for charting the English Channel. While both England and France remained reliant on Dutch charts into the seventeenth century, by the end of that century both had produced their first national maritime atlases. France's *Le Neptune François* (1693) was more favorably viewed as a "superior performance"—by the Dutch and even the English—over England's *Coasting Pilot* (1693) thanks to the ability of the French engineers to benefit from the triangulation network created by the measurements and observations of the French astronomers and surveyors working along the French coastline (see chapter 2). Such works would lay the foundation for subsequent charts as both countries not only hoped to better chart the Channel and their coasts to protect their merchant shipping and their vulnerable coastal sites, but also to challenge one another for control of the Channel and to launch potential invasion plans.

The eighteenth century would witness both countries often at war with one another in the Channel, on the continent, in their respective colonial interests or some combination of all three. They were often at odds as they each attempted to expand their influence and/or dominance on the continent and in existing or new colonial opportunities. Navigational skill and scientific expertise became more important with this expansion. While chapter 4 will address the mapping and charting interests for the colonial enterprise, this chapter addresses the appeal to scientific authority or rhetoric of scientificity in the charts of the Channel. Moreover, it traces the commercial activity intertwined with chart production in both England and France as it demonstrates both the benefits to chart making and its frustration on account of those state institutions (e.g., Dépôt, Académie, Royal Society, Admiralty, etc.) involved or interested in the improvement to charts. By the last decades of the eighteenth century with improvements to navigational instrumentation (i.e., the sextant and the chronometer), the solution to the problem of longitude (by the lunar distance method or use of the marine chronometer), and the increase in the number of scientific expeditions launched by both countries, Cook's critique of the chart making skills and production practices by compilers, publishers, and seamen began to be addressed. In France, the Dépôt, the hydrographic office of the navy, was re-energized under the expertise of Charles-François Beautemps-Beaupré, a hydrographer with field experience, while in Britain, George III established the Hydrographic Office of the Admiralty in order to improve the safety of navigation for His Majesty's fleet called upon (increasingly) to voyage in unfamiliar seas in the formation of the British Empire. Finally, the chart makers on both sides of the Channel from Greenville Collins and Charles Pène at the end of the seventeenth century to Edmond Halley, Jacques Nicolas Bellin, Thomas Jefferys, and Jean-Baptiste Degaulle in the century that followed, all intended that their scientific work be of service to the state and sought support or patronage from various state entities, but every chart was also a product of commerce and success in the commercial sphere meant promoting one's work as authoritative over another.

SIXTEENTH-CENTURY SAILING DIRECTIONS IN ENGLAND AND FRANCE

For many English and French navigators sailing along the western European coast from the Baltic Sea to the northwestern coast of Africa, written sailing directions (perhaps accompanied with some drawn coastal views or charts), the compass, the log-line, the lead and line, and the method of dead reckoning were sufficient even preferable to marine charts well into the eighteenth century. As a consequence marine charts often remained in manuscript form passed from one experienced pilot to another. Sailing directions dominated the art of navigation, but, as pilots voyaged further from home waters and began (with their respective governments) to compete militarily and economically, chart making grew in importance.

In 1483, Pierre Garcie an experienced pilot from Poitou was the first to compile sailing directions for the region from the English Channel to the Bay of Biscay which would come to form *Le routier de la mer* printed in the early sixteenth century. In the opening page of his longer and more detailed work, *Le grant routtier* (Poitiers, 1520), he acknowledged the perils of the sea and the need for navigators to know compass directions, the winds, the tides, and the role of the moon, the sun, and the North Star in wayfinding as well as the customary laws of the sea.[1] His work was a textual description of the Atlantic coastline. The coastal views from woodcut blocks may have seemed familiar to navigators along this coast but appear quite primitive reflecting their pragmatic and non-aesthetic goals as well as the reliance on written descriptions giving compass directions, depths, and distances to aid sailors to avoid coastal hazards and recognize best approaches to harbors and channels. Garcie wrote that the work was composed with the advice and counsel of all the masters and pilots in the noble, yet very subtle, hazardous, and dangerous art of navigation and the "quasi-divine" maritime profession. In fact, the work is dedicated to his great friend and godson, Pierre Ymbert, who was also a navigator. Perhaps in an initial effort to compile (1483) all the received wisdom he knew of to keep his godson safe, Garcie had created a work that when published (1520) aimed to keep many other mariners safe.

Garcie's *Le grant routtier*, by a pilot for pilots, was well received by his compatriots and continued in several editions to 1643. It even made its way to England being printed by Robert Copland as *The Rutter of the See* in 1528.[2] As with the French version these rutters or sailing directions were used in conjunction with calendars, tide tables, and manuscript charts for the specific area to be sailed.[3] Copland's *Rutter* would continue to serve mariners until the mid-1570s after which the English (and French) translations of Dutch works would dominate the trade (see below). There were, however, English mariners who like Garcie voyaged along Europe's Atlantic coastline and even further afield allowing them to gather navigational information to share with fellow pilots and to expand merchant opportunities.

For example, Roger Barlow (d. 1554), a successful Bristol merchant who had travelled to South America with Sebastian Cabot, then the pilot major at Seville, had invested in this Spanish undertaking (1526) to seek (though unsuccessfully) the spices of the east. Upon his return to England, he harbored a plan to initiate an English route to the Spice Islands.[4] Nearly a decade after his return, Barlow produced a manuscript (1540/41) based primarily on Martin Fernandez de Enciso's 1519 work, *Suma de Geographia*, and concluded his description of the fourth quarter of the world with his proposed northerly route to the Spice Islands. With this vast compilation of sailing directions, Barlow hoped to receive royal support for publication and dedicated his manuscript to Henry VIII: "I humblie present unto your maiestie for that your grace maye comande it to be corrected and amended and to geve your auctoritie to be put fourthe in prynte, for that it maye be agradable to the readers and also profytable to suche as intende to have knolege of the navigation to dyscover straugne countries."[5] At this time, there was just too little interest on the part of the crown to set the "ship of State" at sail to distant shores.[6]

In addition, since Barlow's manuscript did not vary from the navigational knowledge presented in Jean Rotz's *Boke of Idrography* (1542) which had far greater aesthetic appeal, Rotz was more successful in obtaining royal patronage. After being frustrated with the patronage from the French king, François I, Rotz, from the active port city of Dieppe, travelled to England and was taken into service to the English king. Dieppe was the home port for many enterprising merchants who supported overseas ventures in the sixteenth century as well as a group of map and chart makers that served these merchants and interested elites. Rotz was a member of this Dieppe school of cartography whose beautiful manuscript maps were coveted by elites and depicted the contemporary knowledge brought back from many of those overseas ventures.

Gayle Brunelle suggests that Rotz, like many other map makers of the Renaissance period, knew well how to present information to their specific audience: "When it was expedient to assert a patron's claim to possess the world or large portions of it, cartographers produced maps that at once imagined a new world for their patrons, and simultaneously confirmed and validated their patrons' sense of their place in that world—at the helm, of course. . . . [E]arly modern maps were produced for a courtly culture in which political and dynastic rivalries and European cultural superiority were imperative at least as important as science in determining what went into the maps, and what was left out."[7] Over the course of two centuries, while national ambitions were protective of state interests, there was growing concern that charts not only include the most recent information, but that these works present a keen sense of scientific authority. The rhetoric *on* the chart and surrounding the production of the chart came to reveal a participation in the Enlightenment ideology of a well reasoned analysis of sources to produce *certain* knowledge.

In the 1540s, Rotz's masterful manuscript work based on the information gathered by him and other pilots that sailed the waters earned him a post as Henry's royal hydrographer (for a time) and the task to chart the English coasts for purposes of defense.[8] The work was a decorative display of the state of the art of navigation including information such as the Regiment of the North, declination tables of the Sun, direction of the winds, compass bearings, and the means of finding latitude of various locations. Rotz's work was an armchair delight but would not become a staple for the mariners along the Atlantic coast or even the Mediterranean coast as all would continue to rely on sailing directions perhaps supplemented by a manuscript chart drawn by the pilot or drawn for him based on his notes. To the mid-sixteenth century, then, the initiators of the production of charts and sailing directions whether in France or England had come from the merchant/mariner community rather than directives from their respective crowns or leading government officials.[9]

From the perspective of the monarch, however, there was an immediate concern regarding coastal fortifications. Henry VIII and François I, had concerns about the vulnerability of their coasts and, initially, brought Italian engineers in the first half of the sixteenth century to work on coastal fortification plans. They were, however, just that: plans for the protection of the land but not more detailed coastal surveys for navigators. Attention to the art of navigation post-Garcie was still in its initial stages.

In France, the mathematician and astronomer, Oronce Finé, a client of François I, held the chair of mathematics at the Collège Royale, founded by his patron. His work on map projections and determining longitude by use of lunar eclipses participated in the discourse on the art of pilotage. Nicolas de Nicolay, royal geographer, translated Pedro de Medina's navigational manual, *L'Art de Naviguer* in 1554 and dedicated it to Henri II. From his own travel experiences on land and on sea, Nicolay wrote that he found the work very useful and necessary for all voyagers. Those who undertook their travels by the authority of this great king would be well served by such a work that would aid in the expansion of his dominion and, like Phoebus, bring light to the universe. France, however, soon found itself embroiled in the Wars of Religion, and Phoebus's sun chariot was halted in flight.[10]

In England, however, while there was a similar shift to native talent by the 1560s, there was no longer dependence "on military engineers, or indeed on its own official servants, for the provision of maps and plans, even in the military and defense spheres."[11] Talented mathematical practitioners, instrument makers, cartographers, and surveyors filled map demands for the crown, ministers, landed elites, and merchants. Stephen Borough, the Master of Trinity House at Deptford Strand, urged Richard Eden, translator and former secretary to Lord Burghley, William Cecil, to translate Martín Cortés' work which would become in English the *Art of Navigation* (1561).[12] Susan Rose suggests that Eden's translation marks "a decisive point in English attitudes to navigation . . . [which] was to be based in the future not on experience and intuition alone but also on astronomical and mathematical principles."[13] In the second half of the sixteenth century Englishmen began setting the foundation for a knowledge base to match the growing propaganda to project England to its logical position as a maritime powerhouse.

In 1574, William Bourne, mathematician and former navy gunner, produced a more popular (and user friendly) manual, *The Regiment for the Sea.*

With log and line and dead reckoning pilots could keep track of their progress on sea charts and add information to their rutters.[14] The stage was soon set for Drake's first circumnavigation of the world 1577–1580, as was a bold challenge to Spanish power; Elizabeth's aid to the Dutch rebels set in motion events that would lead to the Spanish Armada. In 1588 just after this momentous event, physician and mathematician, Thomas Hood began public lectures on the usefulness of mathematics in particular to astronomy, surveying and navigation.[15] In his inaugural address he reflected the sentiment of his fellow Elizabethans, such as Dee, Hakluyt, and others, to challenge the dominance of Spain.[16] Hood delivered this address in the home of Thomas Smith, a merchant leading the way for these lectures and the future head of the English East India Company.[17] Hood's lectures were supported by the Privy Council and funded by the City of London and addressed navigation and hydrography as well as the use of several instruments to aid in astronomical observations to serve navigators.[18] Here, then, is evidence that near the end of the sixteenth century while there was still motivation from the merchant corporations to improve the art of navigation, there was clearly support from the government to present and disseminate this knowledge.

At this point in the late sixteenth century, while popular works to aid pilots in the use of charts were available, many French pilots, like English ones, continued to use the sailing directions published in the early sixteenth century and amended them with up dated information. Similarly, their engineers working along the Atlantic coast (and the eastern and northern interior boundaries) to establish or repair fortifications revealed an interest (on the part of their respective governments) in serving maritime activity, but the coastal fortification plans did not serve the pilots entering the region who still relied on traditional rutters. Moreover, both England and France were extending their maritime ventures beyond the long Atlantic coast. Jacques Cartier voyaged to Canada in the first half of the sixteenth century and Samuel de Champlain in the early seventeenth century. In addition to establishing the Muscovy Company in the mid sixteenth century, voyagers such as Martin Frobischer travelled to Canada, and Francis Drake and Thomas Cavendish conducted their voyages of circumnavigation in the second half of the sixteenth century.[19]

French and English adventurers began to challenge the Iberian dominance, in particular the Portuguese spice trade, and while much knowledge was gained and new sailing directions produced, neither the French nor the English established a printed chart trade or any organized effort to provide pilots with charts. While England was poised with an active merchant body, a growing knowledgeable piloting body, and a government motivated to challenge (and surpass) the Iberians in their overseas ventures, Kenneth Andrews suggests that the Elizabethan Crown "welcomed the efforts of its subjects to develop oceanic enterprise, but it always treated them as secondary to the main issues of European power politics, leaving the initiative—especially the financial initiative—to others."[20] France, on the other hand, would need to wait for the Bourbons to take firm control and initiate navigational projects and overseas trading ventures. It would take, then, the Dutch, pre-eminent at this stage in the late sixteenth century in their trading ventures, to combine the traditional sailing directions, the new navigational manuals, and charts for a comprehensive work to be carried on board.

SIXTEENTH-CENTURY DUTCH SEA ATLASES

As the Dutch were initially more successful in challenging the Iberian voyagers to the South, they were the first to combine sailing directions, much improved coastal profiles, and charts (and a navigational manual) into one work. Lucas Janszoon Waghenaer's (c. 1533–1606) work, *Spieghel der Zeevaerdt*, provided charts with markings of bearings, distances, water depths, currents, anchorages, hazards, and coastal profiles, as well as the textual information on safely sailing along the coast and paying attention to direction, prominent coastal markers, currents, winds, times of tides, and details of the harbor and the bottom.[21] In what would become a customary call by chartmakers for information, Waghenaer, from Enkhuizen—one of the Dutch trading centers—testified to the efforts he had made in constructing his charts, but sought up-to-date information to revise and correct them; "it is quite possible that some faults may be found in this book, which the industrious seaman can nevertheless easily correct on these charts with the help of good sources, for it is less difficult to improve that which is found then [sic] to find something new."[22] Waghenaer's work, published by Christophe Plantin at Leiden in two volumes of 1584 and 1585, covered the western coastline from the Baltic to Cadiz in 45 charts each with the corresponding sailing directions on the reverse.[23] The *Spieghel* included an introduction on the usefulness of charts as well as the material one would find in navigational manuals of the era, that is, lunar tables, tables for the declination of the sun, table of fixed stars, tide tables, and methods for making astronomical observations for calculating latitude.[24]

This work of unique compilation acquired impressive patronage as it was accorded printing rights by Philip II of Spain at the end of 1579 and in early 1580 by the States of Holland. The work, presented to the current stadtholder, Prince William of Orange (1533–1584), was received with "pleasure and favour."[25] Waghenaer's patron François Maelson (1538–1602), a pensionary of Holland who was to be sent to England (in 1585) as ambassador, won for Waghenaer a remuneration from the States of 600 pounds and brought the work to the attention of Elizabeth's Privy Council.[26] England had sent military support to the Netherlands in 1585 to aid the Dutch in their uprising against Habsburg rule, hence Waghenaer's already prepared extravagant dedication to Elizabeth I was easily slipped into the Latin edition, *Speculum Nauticum*, which followed in 1586. Lord Howard (1536–1624), a member of Elizabeth's Privy Council and the Lord High Admiral of England, was pleased to support the Latin edition, but English seamen required a work in English. In 1587 the same year that Sir Francis Drake raided the Iberian coasts and destroyed vessels in the harbor of Cadiz, Anthony Ashley (1551–1627), the clerk of the Privy Council, was given the task of translation by his friend and patron, the lord chancellor, Sir Christopher Hatton also of the Privy Council. Ashley translated the work for an English audience and dedicated *The Mariners' Mirrour* to Hatton. (A French edition appeared in 1590.) These works were delights for the eyes, instructional for the armchair pilot, and attractive to booklovers, but they were not of good utility for workaday mariners.[27]

Despite the approbation of three pilots for the publication of the *Spieghel*, it was not universally praised by mariners who much preferred Waghenaer's larger three-volume work in 250 pages, *Thresoor der Zeevaert* (1592). This work, extended to cover the Mediterranean Sea, included half the charts, but more coastal profiles. Mariners simply found the traditional style of manuals with rutters sufficient. Waghenaer still sought powerful patronage as he dedicated this new work to Prince Maurice of Orange, the new Stadholder of Holland.[28] Six years later, navigators were drawn to Waghenaer's pocket-sized rutter, *Enchuyser Zeecaertboeck* (1598), which returned to the detailed sailing directions and coastal profiles without charts.[29] While Waghenaer's charts were mere sketches, he was the first to place them in a pilot guide, and while his *Spieghel* was too elaborate and expensive for the working mariner, he set a standard for future sea atlases. For much of the seventeenth century, English and French pilots would be served by Dutch charts and atlases or their translations. The path to a national maritime atlas produced in England or France, however, began over concerns for coastal fortifications and defense in the English Channel.

COLBERTIAN MARITIME PROJECT

By the early seventeenth century the French government had taken direct control of the kingdom's fortifications and required the establishment of the post of a royal engineer in the strategic eastern provinces, a practice that would not only grow in numbers but also to include all the provinces over the course of the century.[30] These royal engineers who were increasingly French succeeded their Italian predecessors and produced manuscript (and printed) maps (and charts) of various strategic locations in France.[31] Although much of their cartographic work would remain in manuscript form for obvious security reasons, some works were engraved and became the basis for some early seventeenth-century atlases. Based on these works as well as his own, Christophe Tassin, a royal engineer-geographer and uniquely a seller of prints, published in 1634 an atlas of the coasts of France, *Cartes générale[s] et particulières de toutes les costes de France.*[32] He dedicated this work to the superintendant of the Marine and the king's first minister, Cardinal Richelieu, who may very well have initiated the production of the work, having recognized the necessity of strong coastal fortifications and the development of France's naval presence.[33] At this point in French map/chart production, however, the focus was terrestrial. These coastal maps were aimed more at defense or had a terrestrial origination rather than a hydrographical nature.[34]

There is no doubt that Richelieu supported the development of maritime works, even the construction of an Atlantic fleet (the Ponant) and the financing of a domestic shipbuilding industry, but France always had to be mindful of its continental neighbors and enemies, hence terrestrial fortifications and the army took up considerable State attention, and when pressed the State would direct its financial strength inland (see below). Richelieu's support for the Ponant did not succeed him.[35] In fact, in contrast to England whose naval affairs continued to develop during its Civil War (1642–1651), the internal disruption of the Fronde (1648–1653) weakened naval development and any attention to the Ponant was replaced by the traditional maritime focus on the Mediterranean (the Levant) and the Iberian coastline.[36]

By the 1660s, however, the policies of Colbert, Louis' controller general of finances, secretary of state for the navy and the royal household, aimed to promote the *gloire* of Louis XIV, which for the ruling monarch of the Bourbon dynasty meant territorial gain and tangible recognition of his successes for the day and posterity. The influence of Baconian science and the early stages of Enlightenment encyclopedism led to works that intended to demonstrate a scientific authority. This rhetoric of scientificity whether *on* the chart or in its accompanying report would come to serve less the personal glory of the monarch and more the power of state.

As a consequence of the necessity to build an administrative system to facilitate the opportunities to acquire such renown (initially, the glory of the king and soon, the state) and to then publicize or make known the remarkable circumstances of its acquisition, Colbert set in motion several projects. Among them was the mapping of France. In 1663 Colbert sent instructions to the royal officials in the provinces to gather information for the state, which was to include geographical knowledge. Having found their reports less than adequate, map making became one of the primary concerns of the Académie[37] (see chapter 2). In 1666 perhaps as part of Colbert's campaign to compile information, Charles-Jean Colbert du Terron (1618–1684), *intendant general* of the Atlantic fleet and cousin to Colbert, advised him of the best source for hydrographical information, the Dutch sea atlas, *De Zee-Atlas ofte Water-Wereld*, which had recently been published in Amsterdam by Pieter Goos (1616–1675), engraver, map maker, and map publisher.[38] Colbert asked his librarian to acquire the work for his library. The lavishly colored atlas was quite beautiful and an appropriate adornment for Colbert's library, but not practical for the Atlantic mariner. It is more likely that mariners were using copies from Goos's earlier work, *De Lichtende Columne ofte Zee-Spiegel* (1650), which was more appropriate, that is, less ornate with compass bearings soundings, good anchorage sites, and hazards. Nevertheless, despite the works of talented French engineers along the coasts in the first half of the century, French navigators and state planners had to rely on the Dutch for hydrographic works. Certainly, there were more pilots of the talent of Pierre Garcie who now more than a century later could translate piloting information to charts.

Perhaps in an effort to hasten such developments, a hydrographic school was founded at Dieppe on 30 September 1663 in which Guillaume Denis (d. 1690), a local priest from the church of Saint-Jacques, taught the art of piloting to young men. In 1671, Colbert wrote to Denis praising his work and asking him to send the names of those with knowledge of the Channel coast to him as well as sending him monthly the names of those taking the courses on piloting.[39] War with the Dutch was imminent, and Colbert needed knowledgable mariners. A year later, Colbert wrote to M. D'Herbigny, *Intendant en Mission* (26 August 1672), regarding Denis's complaint that the navy received several seaman with no knowledge at all neither theoretical nor practical in hydrography which caused great inconvenience to the navy.[40] Denis asked that they not be accepted, in effect, until they could pass exams. Colbert asked Herbigny to evaluate this proposal and the consequence of it to the service of the king and commerce.

With regard to making and using marine charts, in a letter of 20 April 1670, Colbert wrote to Terron that the French needed to think about making maritime maps of all their voyages so as to free themselves from a reliance on the Dutch and to use the information that pilots gathered to make maps more accurate.[41] In August of the same year (1670), Colbert emphasized that care should be taken to draw from all the navigational information and pilot journals not only for those who would make the same voyages, but also for composing maritime maps.[42] This letter opened with a comment on Captain Machaut who had abandoned a vessel trading with Guinée. While the captain was not punished on that occasion, Colbert wanted the commanders of these escort vessels to know that they were either to perish or return with the merchant vessel. Abandonment of one's post was not to be tolerated. Such a vessel abandoned might be subject to enemy capture or lost at sea and valuable navigational information not to speak of any merchant goods might be lost or fall into enemy hands. As England and France would come to challenge the Dutch hegemony of overseas trade, ships' journals and sea charts became valuable commodities and while everyone was interested in more accurate charts, the idea was to benefit one's home country not competitors in the marketplace.

For nearly a century the Dutch dominated the commerce of the seas, and England and France had only recently developed a fleet or strategy (i.e., privateers) to attack Dutch maritime affairs and even one another. Begrudgingly political and economic alliances were often made. For example, in June 1670 Louis XIV and Charles II signed the Treaty of Dover, a secret pact for Charles to declare himself a Catholic and break off his alliance with the Dutch. In return for a hefty subsidy and troops to put down the inevitable civil disturbance in England, he would join France in the coming war against the Dutch. While Charles faced his challenges at home and the reality of his pact with the bellicose French king, Louis moved closer to war with the Dutch.

In November 1670, Colbert—several months before he wrote directly to Denis—wrote to his cousin and suggested that given the fighting that could go on in the English Channel, Terron should seek out experienced captains and pilots who could provide pertinent information on the tides, the currents, the winds, the movement of the water along the land and the capes, even provide the tables of the movement of the water. Furthermore, he asked Terron to judge if the *ingénieur* (engineer), La Favollière, could accomplish the kind of work Colbert required.[43] This project was not about fortification but about the movement of the seas and demanded work of a hydrographical nature, and if it were possible, Colbert intended to duplicate the project along France's other strategic coastal sites. La Favollière had the support of Terron, and in 1674 Colbert wrote that his work was not only in service to the king but for the good of the commerce of his subjects.[44] The arrangements, however, for these very useful undertakings could be quite challenging not only because of the nature of the work and the challenges of the terrain, but the upfront expense that could be born by the *ingénieur*. Louis-Nicolas de Clerville (1610–1677), commissioner general of fortifications, was reassured by Colbert that he not only recognized the quality of his work but also the expenses incurred. He promised the *ingénieur* that when he showed de Clerville's work to the king, he would make sure to address the issue of compensation.[45]

Since he had become secretary of state for the navy, Colbert had been grooming his son, Jean-Baptiste Antoine Colbert, marquis de Seignelay (1651–1690), to take his place, and Seignelay had already paid for the right to succeed to the office. On 24 September 1671, Colbert provided his son with a mémoire (or instructions) to guide his actions among which would be the development of the navy. The king, Colbert reported, desired an exact description of the coastlines of his kingdom, and these maritime maps were to be drawn from the navigational information and pilots' journals collected in all the ports of the kingdom.[46] Just a few years later, in the midst of the Dutch War, when the alliance with England was looking quite fragile, Colbert further instructed his son that it was his task to continually examine the maritime maps and assess all the possible scenarios that could arise. Perhaps as a means to that end, Colbert in 1676 charged his son with the duty to collect and inventory all the pertinent maps and to have them copied in the same scale with accompanying reports by the *ingénieurs* who produced the charts. Although by 1674 on account of the expense, Louis had withdrawn his Atlantic fleet to their harbors, over the course of the 1670s this team of *ingénieurs* including the likes of La Favollière, La Voye, Du Bocage, Sainte-Colombe worked independently of the Cassini-Picard geodesic program and produced maps of the French coastline from Brittany to the border of Spain.[47]

For example, in 1678—nearly a year and half (19 September 1676) after Charles Pène, a royal *ingénieur*, was commissioned to complete surveying work to be included in an atlas for the king—Colbert instructed him to draw maps (also to be part of this atlas) of the coast of Normandy. These "exact" maps were not only for the use of French navigators but also a means of ensuring that the coastlines were safe from the enemy.[48] The king would write to the governors and the lieutenants in Normandy and instruct them to provide Pène with all the assistance that he needed to carry out the king's instructions.[49] He requested the production of hydrographic charts as well as the collection of information from all voyages in order to locate the best and safest anchorages and harbors for the naval and merchant fleets. Three years later (1681) Colbert made official what he had suggested in practice for about a decade that pilots turn in their ships' journal to the clerk of the admiralty so that important information could be gathered by the naval ministry.[50] The work of Pène (later *hydrographe de la marine*, hydrographer of the navy, 1691–1701) and other *ingénieurs* at the direction of the state; the state's attention to the collection of shipboard journals and the reports from experienced pilots; and its intention to produce charts in the same scale all assisted in the production of charts compiled to form France's first maritime atlas, *Le Neptune François* (1693), made by order of the king and printed at the *Imprimerie Royale*.[51] As a source of hydrographic information for royal and merchant vessels leaving and returning to the French coast it was an early work of the Colonial Machine which both directed and supported the efforts to create and publish the work. Once completed, however, Joseph Sauveur and Charles Pène sought a subvention from the king given the expense to draw, engrave, and print the atlas.

SAMUEL PEPYS AND THE "CARE OF THE SEA"

In England, Samuel Pepys (1633–1703), the secretary of the navy, was in a unique position to comment on the state of seventeenth-century hydrographic activity.[52] He expressed his frustrations at the slow progress and urged the Crown to take the lead. He was well aware of the lack of patronage for mathematics and the sciences and the decline of such practices as map making and globe making over the reign of Charles I and during the Interregnum.[53] The early years of the Restoration (beginning 1660), however, witnessed great optimism (see chapter 1), and, with a stable and expanding economy and a stimulated intellectual environment, the pursuit of the arts and sciences not only benefited, but also the influential gentry class and a growing middling sort supported a vibrant commercial sphere.[54]

Nevertheless, progress on the independence from foreign works (mostly Dutch) was slow. In 1680, he lamented on the publishing situation of maritime charts noting that Anthony Ashley's translation of the Dutch Waghenaer work in 1588 was not followed by more than one subsequent edition (1605), and England had lapsed back to relying on Dutch works. Although he was quite pleased with the globes Joseph Moxon (1627–1691) made, purchasing one for the Admiralty as well as one for his personal use, he made no mention of Moxon's *A Book of Sea-plats*, a work of six charts of European waters, published in 1657. From Moxon's petition for the position of royal hydrographer (1661), it is apparent that he intended to produce or republish a necessary maritime work and hoped that royal approbation would offset the financial burden.[55] Moxon was appointed royal hydrographer and did republish his *Sea-plats* in 1665, but it was merely the same charts from the first edition with little change. Moreover, the charts were copied from the Dutch work of Pieter Goos. It is, therefore, not surprising that Pepys made no mention of any "exact and perfect Waggoner" by Moxon. As Sarah Tyacke notes until the 1670s, the English relied on manuscript charts or translations of Dutch works.[56]

Pepys commented on the on-going project of John Seller, a royal hydrographer and Moxon's contemporary, but there was little to celebrate there as well. Pepys wrote: "Seller's maps are at the best but copies of the Dutch, with such improvements as he could make therein by private advice upon the observations of single men."[57] Seller's *English Pilot*, volume one, of 1671 was primarily drawn from old Dutch plates with little uniform updating. He had won a thirty-year privilege for his planned *English Pilot* series in five volumes. In the dedication for volume one to the then Duke of York and Lord High Admiral, the future James II, Seller wrote: "'Tis no small disparagement (Most Illustrious Prince) to this Famous Nation, that strangers [i.e., the Dutch] should deprive us, not only of the honour, but advantage of setting forth general descriptions of the Sea-coasts. . . . [A]s if the knowledge of

English Men in the Art and Practice of Navigation were so imperfect, or our Atchievements [sic] and Discoveries abroad so inconsiderable and inferior to theirs, that we must see no further than their books direct us."[58] Writing just four years before Ogilby published his *Britannia* (1675), Seller expressed a similar desire to exhibit the skill and expertise of Englishmen. No more would their countrymen "truckle under the Fame of France and Belgium"[59] (see chapter 1). Seller would humbly take upon "this great Trouble and Charge, which hath been the first Essay of this Nature that hath been done in England."[60] And yet, Seller had to admit that he would draw on the works of those "strangers." Pepys and others recognized that the promise of a truly English Waggoner had yet to be fulfilled.

Given the optimism in the early years of the Restoration, the efforts of the gentry in the Commons to take a leadership role in government, and Pepys' leadership at the Admiralty, there was hope that a stronger directing hand—not necessarily a Colonial Machine—would emerge for the safety of navigation.[61] Pepys was troubled that there was so little sustained effort for improving maritime activity. "[T]he care of the sea has ever been so little in esteem among us that there is no provision in our Parliaments for any standing Committee for the same, as there are for several other things that cannot be thought of greater import to this nation as such."[62] Pepys certainly would have been aware of the efforts, which came from the government and from the merchant community, to better train pilots and officers, but continually advocated for a sustained directing role by the crown (see below).

After all nearly a century before, Richard Hakluyt in his *Principal Navigations* (1598) had encouraged the instructions of sailors and explorers in navigation. In the sixteenth century, Henry VIII established a Trinity House at Deptford, Hull, and Newcastle-on-Tyne for the purposing of educating navigators, a duty which would continue and expand in the eighteenth century. As mentioned earlier, under Elizabeth I, the Privy Council supported (and the City of London funded) Thomas Hood's public lectures in mathematics in which he addressed navigation and hydrography as well as the use of several instruments to aid in astronomical observation to serve navigators.[63] Lectures in practical mathematics continued at Gresham College (founded 1598) by the likes of Edmund Gunter, professor of astronomy from 1619 to 1626. Furthermore, in an effort to maintain a consistent level of instruction rather than relying on public lectures or the training of one professor, in 1673, a school was established by Charles II who provided the support for mathematical classes to be taught at Christ's Hospital to forty boys, that is, the first navigational school. Upon completion of training, the Master of Trinity House evaluated the readiness of the boys at the age of sixteen to be apprenticed to a ship's captain. And before Samuel Pepys' death, he would have witnessed the founding of another school by William and Mary in 1694. By royal charter (25 October 1694) the Royal Hospital School in Greenwich (the hospital itself founded to take care of disabled or veteran mariners, widows, and children of seamen) was established to provide boys with a nautical education suitable for them to become navigators and officers in the royal and merchant fleets.[64]

These training programs aside, for Pepys "care of the sea" or beneficial and valuable works for navigators would only truly emerge if the Crown led the way. "'Tis fit to consider the unreasonableness of expecting that any but a *prince should ever beat the charge* which is necessary for the taking good description of places and making of true sea-charts."[65] Crown patronage, then, of Greenville Collins's coastal surveying work would have the potential to launch an independent London chart trade.[66] It was not to be the case as the French production, *Le Neptune François*, was widely recognized as superior to Collins's "ill performance."[67] Pepys described the *Neptune* as "containing that most august set of sea-charts, and particularly that most accurate one of their own coasts, made by special survey and incomparably beyond anything before extant,"[68] and perhaps noted with envy and some irony, "how soon the Dutch have copied the French *Neptune* and not his [Collins's *Coasting Pilot*]."[69] While Collins's work remained in use into the mid-eighteenth century, it was overshadowed by the French *Neptune* whose charts—benefiting from the triangulation network laid down by Jean Dominique Cassini and other astronomers working at the Paris Observatory—were considered superior.[70] By the late 1680s, both countries had completed the hydrographic surveying work that would form the foundation for their first maritime atlases, but, at the end of the seventeenth century, it was France that began to step out of the shadow of the Dutch map/chart trade, and they did so as a consequence of state support and state direction.

ACTIONS TAKEN IN THE CHANNEL

While, in the last decade of the seventeenth century, hydrographically France was just poised to declare its independence from Dutch charts, its naval power, continental military prowess, and ability to conduct hydrographic work further from its coastline was in great doubt. In the 1680s, France had devoted much of its maritime attention to building a strong naval and merchant presence in the Mediterranean so that in the Channel and the Atlantic, its fleet action was limited, relying more on privateers.[71] Faced with the Glorious Revolution in England, Louis XIV had hoped that the Dutch invasion would throw England into civil war and allow him freer reign for his continental military activities. He had miscalculated. An upset in the balance of power occurred as the English Parliament invited James II's protestant daughter, Mary, and her protestant husband, William of Orange, the Stadholder of Holland and enemy of France, to sit on the English throne. Louis had already declared war on Emperor Leopold II ushering in the Nine Years' War (or War of the League of Augsburg) and William was only too happy to ally with the Emperor against the French king. James II looked to Louis to support him in his attempt to take back the throne.

By late summer 1689 it was clear that the naval forces of England and France were of equal strength, however, the French forces were operating near or at the limit of its resources.[72] Apparently without the interests of James II in mind, the marquis de Seignelay (Colbert's son) and his protégé, Anne Hilarion de Constantin, comte de Tourville (1642–1701) began to construct a strategy late in 1689 that would see a French fleet in control of the Channel and thus the entrance to the North Sea and free to raid the coast of England.[73] Leaving nearly two months later than intended, in June 1690 Tourville with seventy-six ships of the line initiated a fleet action against Arthur Herbert, earl of Torrington (1648–1716), in command of fifty-six English and Dutch vessels forming the line of battle. At the end of June the naval action which would come to be called the Battle of Beachy Head (or Béveziers) resulted in a French victory and caused great alarm as this French fleet was free to sail the Channel.[74]

There seemed, however, on the part of France to be no plan in place beyond the naval action at Béveziers. The intention of putting to sea before the allied forces could gather never happened leaving Tourville with a broken grand design, that is, no "French naval demonstration in the North Sea" to strike fear in Londoners and the disunited allied fleet, no galley fleet descending upon the English ports to destroy shipping, no ships maneuvering to interrupt coal traffic between Newcastle and London, no two dozen ships to prevent William from landing in Ireland, and so on.[75] The French victory at Beachy Head with little to no follow through was simply not enough to weaken William's postion. Instead, William's victory over James II at the Battle of the Boyne secured William and Mary on the throne. While Tourville was criticized for not capitalizing on this French victory, supply shortages, manpower shortages, poor communication—what Donald Pilgrim calls "procedural and administrative weaknesses peculiar to the navy"—and a lack of "pilots for the shallow Channel waters," in particular, reminds one not only of the importance of hydrographical knowledge, in this particular case, of the Channel, but also the crucial role of the state in the creation and support of an institution to secure such knowledge and implement it.[76]

Louis' department of glory having few victories to celebrate in recent years, commemorated the French victory at Beachy Head with a medal struck in 1690 that displayed the bust of Louis XIV on one side and Louis as Neptune riding a chariot ruling the seas on the obverse. It was a dramatic image from the Sun king's earlier days and reflected a moment of glory that was, as moments are, fleeting. Louis' invasion attempt in the summer of 1692 highlighted the challenges France faced in confronting other naval powers. While Brest had a sufficient harbor for some of the fleet, the absence of other sufficient harbors for a naval fleet along the Atlantic and Channel coasts meant that Brest had to rely on the major fleet harbor in the Mediterranean Sea, Toulon. Together, about 1,500 miles apart, these two harbors housed France's naval fleet. Joining them for a fleet action such as in 1692 often relied on the luck of the weather. Moreover, orders or changes to orders from Paris could take days or weeks.[77] In May 1692, Tourville set out without the Toulon fleet under d'Estrées and taking his initial orders, to engage the enemy, to heart, his forty-four ships of the line faced Russell's eighty-four ships of the line off Barfleur. With serious damage much of the fleet broke off in the evening and harbored at La Hogue where the enemy attacked a few days later and destroyed nearly half of the fleet. This time the medals struck for the English and the Dutch revealed a surprised Louis riding a chariot drawn by frogs as the English and Dutch officers *swept* his fleet from the seas; the French cock fled before the unicorn and lion now in possession of Neptune's trident.

CHARTING THE CHANNEL

Back in the Channel, England had missed the opportunity to prevent Tourville from embarking from Brest on his convoy raiding in 1693, but late in the same year and into the following naval season English ships bombarded several French ports from Dunkirk to Brest in the hopes of disrupting French privateering so damaging to the English economy.[78] Pepys, however, lamented the inability or lack of thoughtfulness on the part of the English navy "with the best fleet that ever England was mistress of" to make a descent upon France whose coast was "left wholly open to us." At the same time, once again, he praised the French ability to depict their coasts accurately in the *Neptune François*, "that most august set of sea-charts."[79]

While the scale and detail of the charts of this *Neptune* would not replace the knowledge of local pilots often brought on board to guide vessels into ports, the French success in producing a work praised over the English work was another sort of victory. Both works covered the waterways over which England and France hoped to dominate in order to control the actions (or force the inactions) of the opposing fleet and disrupt trade for its own benefit. Even after both sides gave up their plans for invasion, the arrivals and departures along the Atlantic and Channel coasts remained strategic operations and prompted the publication of charts of this waterway.

Despite Pepys' comment on the quality of the *Neptune* and the Dutch publishers' choice to copy and market it over the contemporary English work, French mariners were not so enamored of it. The work, having been printed at the *Imprimerie Royale*, Louis Phélypeaux, comte de Pontchartrain

(1643–1727), chancellor and minister of the marine, issued an order that navigators were to use it and even threatened to hold the expense for the *Neptune* from their pay, in effect, forcing them to pay for the work whether they obtained it or not.[80] Despite the infrastructure of participating institutions in the Colonial Machine, when the *Neptune* was printed, Joseph Sauveur (1653–1716), mathematician and member of the Académie, and Pène sought a subvention from the king to recoup some of the expense for creating and publishing the maritime atlas.[81] There was, as yet, no directing entity for hydrographical works. In fact, when Jacques Nicolas Bellin (1703–1772), the hydrographer of the Dépôt, was ordered to re-issue the *Neptune*, he had to rescue it "from the dustpin of history."[82]

In England, Trinity House that had supported Collins's surveying work for *Great Britain's Coasting Pilot* now upon completion viewed it as an "ill performance."[83] Even with the approbation of two pilots the completed work had few supporters. Pepys had even asked other map and chart makers in the business to comment on Collins' work. Philip Lea, a London map maker and seller, commented that the work was published by Richard Mount (1654–1722), not a chart maker, as if to suggest that the choice of a commercial publisher was a further damning characteristic.[84] And yet this work, *Great Britain's Coasting Pilot*, with forty-eight charts, coastal views, and sailing directions continued to be reprinted for nearly a century thanks to the chart publishing business of Mount and Page.

By the early eighteenth century the governnments of England and France had launched hydrographic projects which produced respectively their first maritime atlases. While there is evidence that both works obtained a commercial viability, that is, the editions by Mount and Page and the Dutch copies, if Pepys and Bellin are to be believed, the reception by the community of navigators was lukewarm. From Paris, map and chart production was primarily state directed by royal commission, through the Académie, through the Hydrographic Office established in 1720, or a combination of the three. In covering the strategic Channel coasts for the coming decades much of the work out of London emerged from private initiatives (even when public funds were involved), participated in an on-going discourse found in the periodical literature, and fed a commercial map and chart trade that stagnated offering few updated works. The charts published in London and Paris, however, participated in the dominant rhetoric of science in service to the state.

Immediately on the scene was Edmond Halley (1656–1742), clerk of the Royal Society since 1686. After several years of collecting observations and information from pilots' journals and with the support of the Society, Halley presented to the government his grand project, to observe magnetic variation and study the problem of longitude over the course of a world-wide voyage.[85] Halley had, in effect, created his own instruction when he presented his project to the government.[86] Borrowing from Halley's own proposal, the government instructed him

> to use all possible diligence in observing the Course of the Tides in the Channell of England as well in the mid sea as on both Shores, and to inform your self of the precise times of High and Low Water; of the sett and strength of the Flood and Ebb and how many feet it flows in as many places as may suffice to describe the whole. And where there are irregular and half Tides to be more than ordinarily curious in observing them. You are likewise to take the true barings of the principall head lands of the English Coast one from another, and to continue the meridian as often as conveniently may from side to side of the Channell in order to lay down both coasts truly against one another.[87]

The year before the completion of the chart, Halley posted in the *Philosophical Transactions of the Royal Society*, "An Advertisment necessary for all navigators bound up the Channel of England," in which he warned that on account of the change in the variation of the compass and the incorrect latitude of Lizard Point and the Isles of Scilly, pilots were coming into the English Channel too far north and as a consequence unintentionally "running up" the Bristol Channel.[88] A new chart was necessary for the safety of navigation, and with the support of the Royal Society, the Crown, and the Admiralty, Halley provided it.[89] Upon the completion of his second voyage, he presented to the Society (1701) his chart depicting magnetic variation across the Atlantic Ocean. Alan Cook suggests that with his work, "Halley became one of the founders of modern cartography and his chart the model for all manner of maps and charts," and his work a "notable contribution to geomagnetism and more generally to the representation of geographic data."[90]

Halley, also, may have been asked to survey the French coast as evidenced by the Admiralty's addition to Halley's directions: "And in case dureing your being employed on this Service, any other Matters may Occur unto you the observing and Publishing whereof may tend towards the Security of the Navigation of the Subjects of this Majesty or other Princes trading into the Channell you are to be very careful in takeing notice thereof."[91] An examination of this chart demonstrates its attention to the kind of information navigators on both sides of the Channel would find useful. On the map in addition to the notations for magnetic variation, there is information (more toward the coast of England) across the chart on the quality of the bottom (i.e., white sand; grey sand; course sand with shells; etc.); navigational hazards such as semi-submerged rocks, banks, and shoals; the depths in fathoms; the time of high tide; and the direction of the tide. Inset maps include those for Falmouth Harbor and Plymouth Harbor—both "laid down from a survey"—and the Isle of Wight with the adjacent coastline, sandbanks and soundings according to an accurate survey. Such a map as this would have pleased Pepys in the final years of his life, as it participated in the scientific discourse, a bold presentation of the Enlightenment epistemology which joined empiricism and rationalism.

At this time in England, however, the publication of the map was immediately taken over by the commercial firm of Mount and Page. This practice enabled the dissemination of such works, but likely assured that there would be little to no updating of the chart. While Halley had enjoyed important patronage and his work had the approbation of the crown, the Society, and the Admiralty, there was no bureau within the Admiralty with the responsibility of evaluating and publishing charts and overseeing the process of up dating them. At best the Admiralty accumulated foreign and domestic maps, charts, journals, and the like, but there was no administrative procedure beyond the collection. In the early eighteenth century, the practice in England was not too different from what was happening in France for which Halley's chart was a valuable foreign work to purchase and/or copy. It is not surprising that many, if not all, of the subsequent editions in the first half of the eighteenth century by Mount and Page found their way to the Paris archives.[92]

Halley's work would have been among the number of charts in the *Archives de la Marine*, housed in the monastery of the Petit Pères at the Place des Victoires. This early archival collection included several portfolios with documents, plans, maps, and charts but no sense of organization or preparation for collecting or using these documents.[93] Unlike in England, this situation would change as in 1720 under the regency government for Louis XV, the crown established (initially in the same location) the Hydrographic Office or *Dépôt des Cartes, Plans et Journaux de la Marine* (hereafter, Dépôt) not only with the intention that the state benefit from the collection already in existence, but also to put in place a system to gather more sources and to conserve this maritime information for the benefit of the state. Bellin, *ingénieur-hydrographe* (hydrographical engineer), was appointed clerk of the Dépôt and spent over a decade organizing, acquiring, compiling, and copying information from foreign sources in order to produce works that served the royal vessels.[94] Unlike Halley, Bellin never proposed or conducted any scientific projects. As a *géographe/hydrographe de cabinet* (geographer/hydrographer of the study), he operated on the information gathered by others as part of the French Colonial Machine. Bellin's work would be criticized later in the century for perhaps his inability to fully participate in the "reasoned arguments" of the Enlightenment ideology, such that the work of the bureau hydrographer or compiler was viewed inferior to those pilots who had the ability to draw charts[95] (see chapter 5).

Beginning in 1737, Jean-Frédéric Phélypeaux, comte de Maurepas (1701–1781), directed Bellin to make hydrographic charts of all the major oceans and coastlines of strategic importance to French naval and merchant activity, including (though delayed) his chart of the English Channel in 1749. By 1739 England was at war with Spain, France would soon ally with Spain, and Europe was about to enter the War of Austrian Succession (1744–1748). Naval activity in the Mediterranean saw a British fleet fail against a Franco-Spanish fleet at the Battle of Toulon in February 1744. Two days later, in a bold move in the Channel, the French launched an abortive descent on the English coast.[96] England would station warships at Plymouth that could not only keep a constant eye on the Western Approaches and foil any French attempts against England in the Channel, but also harass French convoys attempting to get under way to the West Indies. In addition, the virtually unchecked movement of British warships in these Atlantic waters off the coast of France allowed the British to gather soundings in greater detail and even the failed attempt at Lorient in 1746 had allowed the British to collect data along the French coast to Quiberon Bay, the site of a future naval victory for the British.[97]

In 1749 the year following the Treaty of Aix la Chapelle that ended the War of Austrian Succession, the new naval minister, Antoine-Louis Rouillé, comte de Jouy (1689–1761), ordered Bellin to produce a chart of the English Channel to serve the royal fleet.[98] The title cartouche topped by the royal arms not only included the symbols used on the chart to indicate the times of high tide and currents, but also, as a reflection of the scientific nature and quality of his work, Bellin noted those observations of latitude made at sea as well as those locations whose latitude and longitude were observed astronomically. Moreover, he reported that the magnetic variations noted on the chart were based on observations from 1748. On the map, a navigator viewed not only depths along both coasts but also anchorage sites, shoals, and other navigational hazards, but no inset maps detailing any English ports and no description of the quality of the sea bottom as on Halley's chart. Nonetheless, Bellin's chart was useful for consultation on land or on sea.

In the accompanying report, *Observations sur la carte de la Manche*, Bellin admitted that the publication of the chart had been delayed two years which not only placed its origins with the previous naval minister, Maurepas, but also in the midst of the war that had just ended. The delays, Bellin reported, were due to the challenges of producing useful hydrographic works. He had gathered all the measurements, journals, astronomical observations, and reports from various mariners and, as he often did, emphasized that the difficulty of assembling all of the sources necessary to produce the best work had made hydrography a neglected science in France.[99] With attention and, no doubt from his perspective, greater support for his efforts, France could excel at this science. After all, the English and the Dutch, whom he reported had been at it far longer, had not made the progress one would have hoped for. He suggested that their maps were far from the accuracy necessary for safe navigation. In short, Bellin was advocating for more support from his government while expressing some nationalistic pride, as he intimated that the work he produced was of higher quality and more useful. To produce this chart, he emphasized that he had used the best available sources including the measurements and observations of the French royal engineers, the observations of French astronomers, the 1723 edition of the useful (but rather flawed) Collins' *Coasting Pilot*, the 1734 edition of Seller's *English Pilot*, and the reports of commissions from the admiralty from 1698 on the ports of England from Dover to Landsend.[100] Bellin's *Observations* was offered commercially and printed by J. Chardon on rue Galande in Paris, and his chart was available at the business establishments of map and print sellers as well as his own shop then on rue Dauphine près rue la Christine in Paris.[101] The plate for Bellin's chart, however (unlike the work of Halley), remained in the possession of the institution responsible for its creation, that is, the navy's hydrographic office (or the Dépôt).[102]

In England during the first half of the eighteenth century, navigators were served by republications of Halley's chart of the British Channel from the publishing business of Mount and Page. Halley's chart, originating from a government commission, had become a commercial product and in the midst of the official discussions on the problem of longitude, William Whiston (1667–1752), mathematician and astronomer, suggested that an improved survey be conducted. For its efforts to improve the navigation of the Channel, the English, who had established the Board of Longitude in 1714—which only began to meet in 1737—supported this new survey in 1741. At a meeting of the commissioners of the Board of Longitude, 16 January 1741, Whiston, who had been very active in the discussions of the problem of longitude, was awarded 500 pounds to continue his work to survey the British Channel. Whiston had reported that he, with the help of others in his employ, had made several observations along the coast and required more funding to complete the work. The commissioners agreed that Whiston's instruments (and subsequent observations) would be "to the advantage of navigation."[103] The charts title tells it all: *An Exact Trigonometrical Survey of the British Channel from the North Foreland to Scilly Islands and Cape Clear on the Southwest part of Ireland performed in the year of 1741 and 1742 by Mr. John Renshaw under the direction of Mr. William Whiston partly by private benefactors; but principally by the publick money assigned for this purpose by the Commissioners of the Longitude. Being a correction and improvement of Dr. Edmund Halley's chart of the Channel between England and France, first published in the year 1699.* While it is unclear who Whiston's "private benefactors" were, he had accomplished a new survey of the Channel with state support from the Board, but had relied heavily on Halley's map which he had claimed to be improving.

In the lower left corner of the chart, Whiston advertised not only private lessons in mathematics at his home in Davis Street, Grosvenor Square, but also his availability for land surveying and mapping. The published chart was available at two shops on Fleetstreet in London including the shop of John Senex, run by his widow, Mary Senex, since 1740. In addition the chart was available in Bristol—an important port city whose privateers successfully challenged French guns and continued to trade in the Americas—and Portsmouth, an important port for the Royal Navy and the home of the Royal Naval Academy (established in 1733). Similar to Halley's chart, there were four inset maps of the River Thames, Plymouth Sound, Falmouth, the Scilly Islands, and the Isle of Wight and Portsmouth. Within a decade after its printing in 1745, the plates of this new chart were acquired by Mount and Page who began (as they had with the Halley's chart) offering it for sale.[104] The Board was supportive of Whiston's project but did not take possession or retain control of the chart for the purposes of improving it.

While the ultimate completion of the charts of Halley and Whiston emerged from their own initial motivation, both were (to varying degrees) officially funded and directed projects. In this respect there is a parallel with the charts of Bellin; he was, as the appointed clerk of the Dépôt, acting on orders by the naval minister in his choice of project, and his work was obviously officially funded and directed. Where the achievements diverge is the inability or lack of desire on the part of the Royal Society, the Admiralty, or thc Board to keep or retain control of the chart (or plates) for the purposes of improving hydrographic work. The unrestricted commercial trade allowed Mount and Page to republish not only the works of Halley and Whiston, but also Collins' *Coasting Pilot* and Seller's *English Pilot* with little to no regard for updating material. For Bellin, the reports that accompanied many of his works attested to his efforts to continually update works for pilots and officers. His position at the Dépôt, he believed set his work apart from British works as well as those of fellow French geographers/hydrographers, who did not hold official government positions and did not work with the direct observations coming in the from the field or the numerous other sources at the Dépôt, both merely copied what was already in print.

CHARTING THE CHANNEL AND THE SEVEN YEARS' WAR

At the start of the Seven Years' War (1756–1763), Britain was a singular dominant seapower but had to be wary of any hostile actions from a combined naval fleet of, say, French and Spanish ships. Its vessels outnumbered by 2 to 1 or more, French authorities had long known that it could not challenge the British navy but needed to maintain commercial and administrative links with its colonies and be prepared for defensive actions upon departures and arrivals along the Atlantic and Channel coasts.[105] Initially, small squadrons which had since the late seventeenth century been quite successful in their departures along the French coastline, soon required protection to drive away the plethora of British ships (e.g., the privateers of Bristol) issued letters of marque which resulted in a decline of French sea commerce.[106] Britain also initiated a campaign between 1757 and 1758 to shut down French maritime activity along the Atlantic and Channel coastlines in the hopes that Louis XV would be forced to draw military forces away from the continental campaigns in Germany to protect his maritime activity. The descents on the Atlantic coastline of France did not have the intended result, and, in fact, led to another naval disaster for Britain at St. Cast (1758).

In the summer of 1754 Louis XV replaced Rouillé with comte d'Arnouville, Jean-Baptiste de Machault (1701–1794), the controller-general and keeper of the seals. Machault needed to prepare the ministry for war, and, in 1757, Bellin would produce, by his order, a series of charts covering the British Isles as well as a text, *Essai Géographique sur les isles Britanniques*. Bellin wrote that his goal was to provide navigational charts of the coasts of England, Scotland, and Ireland on which a pilot could find where the principal capes and ports were located according to exact latitude and longitude values, the nature of the coastal terrain, and the distances between locations. To make his charts even more useful he provided a *Routier* found in part two of his work. In part one he provided a geographical description of the interior of the countries, and in part three he provided an analysis of the charts he produced which he described as the most useful currently available in France.[107]

He commented that French navigators had long wanted *cartes réduite*, that is, Mercator charts, of the coastlines of the British Isles and that he had taken great care to present not only the hydrographic information (e.g., harbors, depths, currents, hazards), but also information on the interior of the countries.[108] While the maps included in this work were illustrative (for the sailing directions in part two), they were, by their size, not serviceable to navigators. Consequently, in addition to a reduced version on one sheet, Bellin wrote that he produced a fuller presentation of the Isles for navigators in his five-sheet chart which could be used separately or joined together.[109] The first sheet, *Partie Meridionale de l'Angleterre,* of his larger work covered the same area as his 1749 chart on the English Channel, and while he, no doubt, used that map as a source, this sheet included more details, in particular, along the southern English coastline.[110] He included depths, safe anchorage sites, currents, time of high tide, and hazards.

In the *Essai*, Bellin seemed eager to impress upon the reader of the effort that he had taken to produce the text as well as his various hydrographical works noting, in particular, that he had only two latitude and longitude values for London and Oxford which were not sufficient to fix the southern coastline of England. Consequently, he had had to assemble and analyze a great many English sources, printed and manuscript. In addition to the 1723 edition of Collins' *Coasting Pilot* which was limited because it did not provide latitude values for the various Capes and Ports, Bellin utilized the annual English publication, *The Mariners New Calendar*, from which he could obtain latitude and longitude values for principal locations along the various coasts.

In part two of his work addressing the sailing directions, he also advised pilots regarding the way to navigate in the Channel.[111] For example, one should pay attention to the color of the water, navigational hazards such as shoals and rocks; one should note the height of the pole star, compass bearings, depths, the quality of the bottom, and so on. In heading for the coast of France, he reported that after making sure one has passed by Lizard Point (off Cornwall), one can run for the coast of France following the winds but it was not necessary to pass Les Casquets, a group of rocks near Alderney, one of the Channel Islands. Once near these rocks, it was necessary to be careful of the tides which could carry one onto the islands.[112] In part II, chapter five, Bellin reported on the strategic importance of these islands possessed by the English, located in the Channel and conveniently close to the French coastline. He drew information on their location from the observations of French astronomers and the 1755 edition of the English printed map drawn from the survey of Philip Dumaresq (c. 1650–1690) and drawn by Captain Clement Lempriere, late drawing master in the Tower of London.[113] These Islands had even been the subject of a short article in the December 1753 edition of the *London Magazine* in which the author noted the navigational challenges, "the rocks are vast and terrible; the tides rapid and strong, so that here is no still water at any time, as in other parts of the British Channel."[114] Thomas Kitchin (1719–1784), engraver and map maker, provided the map, *A description of the Isles of Jersey, Guernsey, Alderney, and Sark*, that accompanied the article no doubt drawn from his map in the *Small English Atlas* published in London with Thomas Jefferys in 1749 and again in 1751.

That these islands, in particular Jersey, were of strategic interests to the French is evidenced by the several reports, instructions, and descriptions found in the French naval archives (rather than periodical literature) from the years of the war. Emmanuel-Armand de Vignerot du Plessis de Richelieu, duc d'Aiguillon (1720–1788), governor of Brittany (1753) was asked by the court (1756) to evaluate the ability of the Atlantic coastal ports to defend against an English descent and was given instructions to look into a plan for action against the Channel island of Jersey. Two years later a report was written on the usefulness of an enterprise against both Jersey and Guernsey.[115] Included in the government information was a printed work, Bellin's *Remarques sur les isles de Jersey, Grenesey, et Aurigny*, on the Channel Islands that accompanied his charts. He had noted the production of these Mercator maps or charts in his *Routier* found in part two of his *Essai*.[116] As he indicated in his *Remarques*, his efforts had made these charts far superior to any that had been executed prior.

Among his British competitors was Jefferys, Bellin's contemporary and a subject of both his praise and ridicule. Jefferys, engraver and map publisher, came to prominence in the London map trade during the Nine Years' War (c. 1746) when he was commissioned by Edward Cave to produce maps for the *Gentleman's Magazine.*[117] Shortly thereafter, he became geographer to the Prince of Wales (1748) and, later, geographer to the king (1761). Jefferys' work is most known for highlighting British interests and territorial claims in North America most usually against French pretensions (see chapter 4). In 1758 there was a British attempt to invade France with the goal of drawing French forces away from Germany. British naval policy had already forced France to conduct its maritime commerce under protection of squadrons, but its attempts against the French coast were less successful: "The army did not land at Rochefort in 1757, it did little damage at St. Malo in 1758 and met with disaster at St Cast Bay in the same year."[118]

In 1758, Jefferys published a new chart entitled, *The Seat of War on the coast of France . . . drawn from the latest surveys, and regulated by astronomical observations.*[119] The chart extended from Enkhuizen (home of Lucas Waghenaer) along the coast of the Netherlands by the Zuider Zee south to the Straits of Gibraltar and into the Mediterranean Sea east as far as Toulon along the coast of France and east as far as the Gulf of Béjaia in Algiers. The chart carried eight inset plans such as the plan of the bay and city of Cadiz, plan of Dunkirk, plan of St. Sebastian (in the crook of the Bay of Biscay), and the fortifications of Port Mahon in Minorca with principal sites, depths, and hazards. With the exception of the plan of the Harbor of Port Mahon in Minorca (i.e., Battle of Minorca, 1756) and perhaps the two plans for Dunkirk, there were no plans for harbors or strategic locations along the French coast despite the title of the chart. An advertisement on the map indicated that the table of latitude and longitude values were drawn from the astronomical observations of various academies and societies, "established for improving the Sciences; particularly those of London and Paris." Jefferys indicated that the chart and map sources were drawn from those by French geographers, such as, the chart of the Mediterranean by Bremond and Michelot (which was actually first published in the late seventeenth century in Marseilles) and the chart of the Mediterranean (by Bellin) ordered by Maurepas but with no mention of Bellin's name. More detailed British charts of the French coastline were still needed.

The periodical literature of 1758 was full of commentary, criticism, and illustrative maps to guide the reader through the British (and French) actions. For example, in the June 1758 edition of the *Gentleman's Magazine* a map of the coast of Brittany between St. Malo and Cancale Bay accompanied a report of the actions at St. Malo.[120] The approach to the port city from the sea was too dangerous, and along the way one transport was lost "running afoul of a sunken rock" near the isles of Jersey and Sark. In the October 1758 edition of the *London Magazine* three different articles reported or reflected on the "late attempts in France," that is, the events leading to the Battle at St. Cast in early September 1758.[121] Accompanying an article critical of English action was a chart of the English Channel so that "our readers with an accurate chart of the English Channel, and the adjacents coasts of England and France . . . may trace out the late operations agains the enemy, and the position of our fleets and cruizers."[122] The article criticized the actions of the military for trying to re-embark troops "at a place where no proper measures had been taken for their cover and defence."[123] Perhaps a symptom of not knowing the coastline well enough. There was even criticism for landing troops at St. Lunaire's Bay at which there was no chance to reimbark them because the wind blew directly offshore, and hence, the disaster at St. Cast was set into motion.[124] In the 1758 December edition of the *Gentlemen's Magazine* the editors included an excerpt from a "recently published" pamphlet, *A View of the French Coast of the Number and Strength of their Ports in the Channel, Ocean, and Mediterranean* which provided the British readership with a description of the ports along the French coast with the time of high tide.[125] With the excerpt was a map of the coast of France on which readers could locate the sites of recent events. For British readership there was, then, no shortage of commentary either on the events along the Channel coasts during the Seven Years' War or on graphic representations to assist the reader in following the events, but there was no office or institution as yet with the responsibility to provide up-to-date charts for pilots and officers. In fact, for much of the eighteenth century, British pilots and officers were still served by republications of Collins' *Coasting Pilot*, Seller's *English Pilot*, or Halley's map of the Channel all of which were published by Mount and Page.

France continued to have designs on taking back the Channel Islands and invading the English coastline. While it was doubtful that France would be successful in wresting "long-term control of the Channel from the British," such a possibility did "haunt British naval thinking."[126] When Etienne-François, duc de Choiseul (1719–1785), foreign minister, arrived on the scene the naval ministry was in shambles with several lackluster leaders. By early 1759 Choiseul began to conjure up a masterful plan to force British forces closer to home by invading England; he was treading on familiar territory or sailing familiar seas.[127] The planned invasion by France on the English coast was an attempt to force the British to draw their forces closer to home which would, it was hoped, protect French interests in North America and the West Indies. "Choiseul recognized that Britain could be forced to concentrate its resources at home if a credible invasion threat could be mounted."[128] The challenge for France was to gather its forces both naval and land without tipping off the British who could scuttle the plans. France had no adequate launching point for all to gather. "Since the 1690s France had been compelled to attempt this complex channeling of resources whenever she wanted to undertake large-scale operations in the Western Approaches."[129] Choiseul's plan beginning in early 1759 was a bold but tenuous one as the assemblage of ships had to emerge before British forces could attempt to scatter them.

The British knew well the threat from across the Channel, and in the summer of 1759, Rear Admiral Rodney (1718–1792) bombarded the transports in Le Havre that were to carry the troops in the invasion plan. While Vice Admiral Boscawen (1711–1761) watched for movement through the Straits of Gibraltar, Vice Admiral Hawke (1710–1781) sat off the Breton coast, in effect, guarding France's Atlantic ports. Again, in late summer the plan was in danger as Boscawen destroyed two French ships and captured three in Lagos Bay.[130] Choiseul still emboldened (or foolish) moved forward with the invasion. The Battle of Quiberon Bay in late November 1759 signalled the failure of yet another French invasion plan. Hawke was victorious at Quiberon Bay in November 1759, but a letter from the admiral dated 24 November 1759 published in the *Gentleman's Magazine* still reflected on the British lack of knowledge on the French coast. He wrote, "Night was now come, and being on a part of the coast, among islands and shoals, of which we were totally ignorant without a pilot, as was the greatest part of the squadron, and blowing hard on a lee shore, I made the signal to anchor, and came too in 15 fathoms of water. . . . In the night we heard many guns of distress fired, but blowing hard, *want of knowledge of the coast*, and whether they were fired by a friend or an enemy, prevented all means of relief."[131] The editors of the *London Magazine* had even published a map of the coastline of France in the December 1759 edition to accompany Hawke's report.[132]

Perhaps Jefferys hoped to alleviate that limitation of knowledge with his two publications on the region in 1761. With Britain still in the midst of the Seven Years' War, he published a two volume work with accompanying maps and plans of the maritime coast of France, *A Description of the Maritime parts of France . . . illustrated with charts of the Sea Coasts and plans of all the fortified places on it*. In the dedication he clearly revealed his nationalistic pride in Britain's prominence.

> To the king's most excellent majesty. Sire, The Navy of France which, but a few years ago, was so formidable as to threaten Your Majesty's Kingdoms with an Invasion, is now either destroyed by the superior Force of Your Majesty's Fleets, or kept blocked up to rot in the Harbours where the Ships were fitting out. That the ports of this once powerful, but now vanquished, Competitor with the British Flag, . . . might be particularly known to my fellow

> Subjects, was the first Motive that induced me to undertake the elaborate and expensive Work.[133]

The remainder of the book is a description of the coastal and fortified areas of France from the north at Dunkirk, to Agincourt, Dieppe, Barfleur, Brest, Bayonne in the Bay of Biscay, and into the Mediterranean to Toulon, and their history up to the present.

In an evaluation of this work, the reviewer recognized its timeliness and utility.[134] "Our general unacquaintance with the enemy's coast, so apparent when the expeditions against St. Malo's, Rochefort, and St. Cast took place, is alone sufficient to evince the utility of the present design." In addition, the reviewer adopted a rather vexed tone on the state of knowledge of certain members in society.

> To the shame of our nobility and gentry who travel for improvement [i.e., Grand Tour], be it spoken, there was not in the kingdom a person, except one gentleman, an engineer by profession, who had the least idea of the harbor or fortifications of Rochefort, at the time an armament was equipping to annoy the enemy by an attack on that quarter. Even our naval officers, who had picqued themselves the most on a perfect knowledge of the coast, and had recommended themselves to notice, by specious pretensions to _____, [i.e., no name given] proved their ignorance, not only of the soundings but of the fit places for debarkation, the avenues to the town, and the strength and position of the fortifications.[135]

The reviewer's comments reflected the concern (and criticism) found in the reports and articles of the *Gentleman's Magazine* and the *London Magazine*. As for the second volume which contained the engraved works of Jefferys with maps, plans, and charts including one for the English Channel, the reviewer concluded that while the maps were well engraved, any comment on their accuracy would be left to more competent judges.[136] The work was to be a source of information, useful in the planning stages but not necessarily for navigators. That audience would be served by Jefferys' other publication in the same year.

In 1761 Jefferys published the very successful *Le Petit Neptune François: or, the French Coasting Pilot*, a translation (with revisions) of *Petit Flambeau de la Mer* by Georges Boissaye du Bocage (1626–1696). The title of the work is rather long and provided not only a lengthy description of his sources used to amend the French original—that is, the 1756 edition of the *Neptune François*; astronomical observations from the Académie; charts from the *Neptune*, the large map of Cassini de Thury, and the works of Bellin, Michelot and Bremond—but also a description of the information within, "a particular description of the Bays, Roads, Rocks, Sands, Land-Marks, Depths of Water, Bearings, and Distances from Place to Place; the Setting and Flowing of the Tides, on the coast of France with Tables of the Latitude, Longitude, and Tides." This was truly a work for the mariner, and, according to the reviewer, "to every other reader will prove not only void of entertainment, but of utility likewise."[137] "The esteem in which the *Petit Flambeau* has been held by all the mariners of France and Holland, cannot fail of recommending the *[Le Petit] Neptune François* to the British seaman, especially at a time when our fleets are continually stationed upon the coasts of France, and the nation deeply interested in a minute knowledge of whatever can render their situation more secure or commodious."[138] Among the graphical sources utilized to produce this English translation were Bellin's charts of the islands of Guernsey and Jersey and a recent survey of lower Normandy. The work was dedicated to George, Lord Anson (1697–1762), Baron of Soberton, First Lord Commissioner of the Admiralty, and a member of his majesty's most honourable Privy Council. As customary (at least) as far back as Waghenaer's sea atlas, Jefferys asked his readers to report any necessary additions or revisions to the work. "If any Gentleman of the Navy shall favour the Editor with Additions or Corrections of any Kind, he will most gratefully acknowledge the Obligation, and make the necessary alterations in the next edition."[139]

Together these two works provided clear evidence that the British now knew what the French knew regarding their coasts. As the reviewer suggested this work would "prove useful and commodious to all who are not sufficiently acquainted with the French language to peruse the large *Neptune François*; or whose situation and circumstances may render such a purchase inconvenient."[140] Enclosed in the work was a map of France with a table of latitude and longitude values for the principal places along the coast, observations by the French astronomers of the Académie, and a list of soundings (and the quality of the bottom) taken by M. Magin at the order of the French king in 1756 and 1757. The charts, sailing directions, and coastal profiles began in the north along the coast of Flanders and extended into the Mediterranean Sea. In short (though this was a large work), Jefferys placed before those unknowing members of the nobility, the gentry, and the naval officers a work that could now truly reflected the prominence of the British naval fleet.

In the closing year of the Seven Years' War, Bellin published an updated chart of the English Channel, *Carte reduite de La Manche*.[141] While there was more detailed information for both coasts on this new chart, hopes to challenge the British in Channel were over. French finances were in grave condition and Choiseul's failed invasion had left the French navy in a shambles. France still sent out convoys to the West Indies and Canada but Britain was an ever present threat, and French naval power was impotent. France would return again to challenge Britain on the open seas, but the war could not end fast enough. Choiseul, undeterred, continued to develop the infrastructure of the navy, the officers training corps, and the amount of tonnage available for merchant and fleet actions. As before a joint action with Spain could threaten the British, but French participation was "dependant upon the fiscal and domestic security of the state," and that was sorely lacking in 1763.[142] After the Seven Years' War, France's vision for a colonial empire was severely dimmed, but trade relations continued making the Channel still an important waterway and a siren's call for invasion regardless of the likelihood of success.

CHARTING THE CHANNEL IN THE FINAL DECADES OF THE EIGHTEENTH CENTURY

With the first publication of the *Nautical Almanac* (1767), the improvement to the sextant, and the development of the marine chronometer, pilots were better able to solve the problem of longitude either by the lunar distance method (used well into the nineteenth century) or by carrying the chronometer on board (more common by 1850). The expansion of trade and colonial competition required that pilots plot their course on charts. And with the increasing number of scientific expeditions new waters were being charted and more demands were being made on the compilers at home.

In 1770, Adrien-Louis de Bonnières, comte de Guines (later, duc de Guines) (1735–1806) became French ambassador in London (1770–1776). Having travelled in England and Scotland, he wrote a mémoire for the king suggesting an invasion of England and accompanied his report with charts.[143] In the king's council there was great dispute over de Guines' project, and no action was taken, but a few years later in May 1775, he brought his plan to the new king, Louis XVI.[144] These plans were still ineffective because France had not solved its problem of assembling a fleet for a Channel action against the English coast. Furthermore there was concern that greater knowledge of the coast was necessary. In fact, many navigators along the Channel coast had resorted to using foreign charts, in particular Dutch charts, over the charts available from the Dépôt. To remedy the situation, the hydrographic schools were required to use charts from the Dépôt and by a royal decree (5 October 1773), the government ordered that this office would control hydrographic production, extending the 1720 order that created the Hydrographic Office for the conservation of such documents.[145] "[A]ll marine charts, portulans, and instructions necessary for the conduct of vessels—warships as well as merchant ships—be composed, prepared, and published exclusively at the dépôt of His Majesty by persons capable of doing the job well and that these works always be accompanied by printed analyses, with an indication of which authorities support the work."[146] Bellin would have been pleased with this decree though he had in the last few decades of his career come into some criticism regarding his "capability" to analyze the authorities and produce, in particular, charts with no active maritime experience.[147]

The War of American Independence reignited invasion plans, but Bellin had died in March 1772, and into the void at the Dépôt appeared a unique character from the provinces, Jean-Baptiste Degaulle (1732–1810), a cook turned seaman turned merchant pilot, who retired from the sea to Le Havre and established himself as an engineer and map maker.[148] In 1773 just a month before the edict addressing the Dépôt's control over hydrographic productions, he published a chart of the English Channel, *Nouvelle carte réduite de la Manche de Bretagne*, dedicated to commerce with the approbation of the Académie but no acknowledgement from the Dépôt despite his best efforts to obtain such recognition.[149] Prior to a second publication in 1776, the Dépôt, in recognition of Degaulle's expertise in the field, accorded him authority at the entrepôt of Le Havre from which he would benefit from the navigational information from arriving pilots and seamen. He had accompanied his chart with a mémoire, and in the 1778 republication of his chart, still dedicated to commerce, he acknowledged the approbation of the Académie (Paris) and the Académie de Marine (Brest), his royal brevet as an engineer of the marine, and the Parisian address of the shop of J. Dezauche (engraver) where the chart could be purchased.[150] The large format of his chart and the greater detail of the coasts served the pilots of these waters far better than the 1763 chart by Bellin, and he was awarded 450 livres for the expense of engraving the chart and 6000 prints, a testament to the demand and a recognition of how useful this chart would be.[151]

While there would be future invasion plans under Napoleon that might be well served by these works, they were often shelved as attention was turned toward the continent. By the end of the War of American Independence and the looming Revolution, French finances could not entertain Bourbon plans for an invasion across the Channel. Bourbon France simply lacked the naval infrastructural support necessary to carry out a sustained operation for invasion.[152] *La Manche* remained a British-dominated seaway in more ways than one.[153] Moreover, England with its map and chart trade centered in London was poised to take the lead in production, and about two hundred years after Anthony Ashley was ordered to translate Waghenaer's *Spieghel der Zeevaerdt*, foreigners now would turn to copy British maps and charts.

In England prior to his death in 1794 and the formation of the Admiralty's Hydrographic Office in 1795, Robert Sayer (1725–1794) and his partner John Bennett (until 1785)[154] were quite successful in their map and chart publishing business. Their *Channel Pilot* published several times into the nineteenth century initially included twenty-one charts covering from the River Thames south along the French Atlantic coast to the Bay of Biscay. Sayer and Bennett purchased charts from members of the Royal Navy, such as, George Burn and John Stephenson; Trinity House pilots, such as, Joseph

Ross and James Grosvenor; as well as other navigators familiar with the Channel. Susanna Fisher suggests that Sayer and Bennett were successful in the second half of the eighteenth century as the publishing house of Mount and Page was in decline and "new specialist nautical publishers" had not yet begun to produce (enough) work to challenge Sayer.[155] Moreover, Sayer was a savvy promoter of his publication list as he made sure to highlight well-known authorities in the field even if he fibbed a little. For example, a map drawn by James Cook (no relation) and published by Sayer and Bennett in 1779 was soon attributed in a later publication list to Captain James Cook, a play that no doubt would have annoyed Captain Cook, who critiqued the current practice of chart publishers.[156]

In a journal entry dated 7 September 1770, *the real* Captain James Cook, who was in his first voyage and at the time off the coast of New Guinea, reflected on the state of chart making from the vantage point of a skilled navigator and chart maker frustrated with the industry. He found blame for faulty charts by "Compilers and Publishers, who publish to the world the rude Sketches of the Navigator as Accurate surveys, without telling what authority they have for so doing." Without a clear indication of the sources of information, chart users would be less able to judge "where to depend upon the Charts, and where not." A similar sentiment was included in the 5 October 1773 French decree in which the accusation was that "a heap of maps [were] published without such particulars . . . [and] although denuded of material sufficient for their construction [i.e., authorities], are announced with fatuous titles, exaggerated in order to promote sales."[157] Cook even shared the blame for faulty charts with seamen who even when capable of drawing a chart were quite false in its rendering: "I have known them [to] lay down the line of a Coast they have never seen, and put down Soundings where they never have sounded; and, after all, are so fond of their performances as to pass the whole off as Sterling under the Title of a Survey Plan, etc." What was necessary was an individual, an office, or an institution whose responsibility it was to evaluate the works provided by pilots in order to investigate if they were indeed "sterling" or merely defective "rude sketches."

CONCLUSION

The trials and tribulations of balancing substantive and authoritative revisions to charts with the commercial enterprise is highlighted by the lawsuit brought by Sayer and Bennett in 1785 against John Hamilton Moore (1738–1805), a respected teacher of mathematics and navigation, who in entering into the chart making business ran afoul of Sayer (and Bennett) who accused him of plagiarism. Moore, however, had experts on his side, including William Wales—an astronomer who accompanied Cook on his second voyage—who declared Moore's charts more advanced than those of Sayer which were considered "erroneous and dangerous."[158] Moore had converted several of his charts to Mercator projection, an obvious advancement that Sayer neglected. The judge on the case, Lord Mansfield, emphasized that publishers had every right to utilize all the works available and as long as the resulting work was not a "servile copy," a chart with *some* improvements could be published. "Whoever has it in his intention to publish a chart, may take advantage of all prior publications. There is no monopoly here . . . but upon any question of this kind, the jury will decide whether it be a servile imitation or not. If an erroneous chart be made, God forbid it should not be corrected, even in a small degree, so that it thereby becomes more serviceable and useful."[159] Fisher suggests, "This ruling left publishers free to make extensive use of earlier charts, as the smallest modification to the original work protected them from charges of piracy, but must have discouraged investment in expensive surveys which would not be protected by copyright."[160] While Sayer did have the financial ability to commission hydrographic surveys, most publishers did not and had to rely on pilots and seamen selling the works they accomplished perhaps as part of commissions for surveys by the Admiralty and East India Company. Hence the potential problems described by Cook with publishers and seamen producing or drawing charts continued. What was required was an office perhaps of the Admiralty that would direct surveying efforts and privilege specific publishers to produce the authoritative charts drawn from those surveys. This full arrangement was decades away.

Even in Paris with the work of the Dépôt and the decree of 1773, there was concern for the quality of published charts (and maps). A broader arrêt (decree or ruling) not specifically addressing the Dépôt was reissued on 10 June 1786 and required that anyone "wishing to engrave, publish, and sell geographical works of any sort . . . obtain permission from the chancellor or the garde des sceaux [keeper of the seals]."[161] Similar to the earlier 1773 arrêt and Cook's critiqued, the map or chart had to be accompanied by their sources, that is, "their foundations or proofs in support of how they were made must be presented."[162] While in France the chancellor or his office passed along the approval for a work to be published, there had been a parallel move to reorganize, perhaps re-energize the Hydrographic Office beginning in 1773. Not only was the navy to embrace hydrographic surveying in an effort to take a more active role in chart making over mere compilation, but the position once held by Bellin as a bureau geographer/hydrographer was to become also a more active "seafaring" position. Also, the Dépôt was to become a privileged location with outside access by permission only in an effort to lend a greater sense of control, conservancy (i.e., document did not "go walking"), and authority to the work being conducted there.[163] The commercial enterprise remained but as the works of Pedley and Chapuis demonstrate there was in the late eighteenth century a move toward greater professionalization, and the Hydrographic Office was ably led by Charles-François Beautemps-Beaupré (1766–1854) beginning in 1799.

In the last decades of the eighteenth century the best piloting chart available in Britain was the new work of Moore, former naval midshipman turned teacher of mathematics and navigation. Moore's *Practical Navigator* (1772) ran to seventeen editions during his lifetime, and while he did *run afoul* of Sayer upon moving to London, he did establish a successful business in nautical publications and in teaching the nautical arts to prepare young men for the lieutenant's exam of the Royal Navy.[164] These connections served him quite well in obtaining information for drawing new charts or updating old ones, and he was most successful with the sale of his large chart of the English Channel beginning in 1786.[165] Being a necessity for both naval and merchant pilots, Moore's claim to have sold "upwards of 5000 copies" from 1786 to 1792[166] is not unlikely and is similar to the 6000 prints accorded Degaulle for his chart of the Channel in 1778. Moore's career encapsulated to varying degrees the three necessary elements for the field of hydrography to move forward with more authoritative works: knowledge of science of navigation and hydrography, a commercial business devoted to the sale of up dated charts, and connections (however unofficial or indirect they may have been) to supporting institutions for hydrographic activity.

In 1795, George III established the Hydrographic Office of the Admiralty to evaluate and use the documents already collected over several decades to serve the navigational safety of the fleet, in particular, called upon to navigate unfamiliar seas. "The great inconvenience which has constantly been felt by the Officers of Your Majesty's Fleet, especially when ordered abroad, from the want of sufficient information respecting the navigation of those parts of the world to which their services may be directed."[167] In addition they were well aware of the activites of France's Dépôt and had even relied on them. "In looking into the practice of all other countries less dependent certainly than this on the effect of naval operations, we find that considerable establishments have been wisely formed for the purpose of attending to this important object."[168] The Admiralty remained dependent on commercial charts, purchasing them from the London publishers, into the nineteenth century. A more official approbation for certain charts came in the first decade of the nineteenth century, as a committee of the Admiralty drew up (from a total of one thousand charts) a list of two hundred "approved" works "to a greater or lesser extent inaccurate" and half from the commercial sphere. The Admiralty would not become independent of the commercial chart trade until the 1830s.[169]

This investigation has revealed that what was required was talented and skilled chart makers, supportive and organizationally-inclined institutions (or offices), and protection for the livelihood and product of those chart makers. The commercial sphere could not provide encouragement for the production of up-to-date and technically superior works on account of the necessary financial expenditures upfront. France had led the way thanks to its Colonial Machine, but by the late eighteenth and early nineteenth century Britain would move forward with its chart making (though not completely independent of the commercial trade) in service to its Imperial Machine.

NOTES

1. D. W. Waters, *The Rutters of the Sea: The Sailing Directions of Pierre Garcie* (New Haven, CT: Yale University Press, 1967). See Pierre Garcie, *Le grant routtier* (Poitiers, 1521), reproduction, at: BN, http://gallica.bnf.fr/ark:/12148/bpt6k54143h.r=Pierre+Garcie.langEN .

2. Waters, *Rutters*, 4. Also see Charles Dugast-Matifeux, "Notice sur Pierre Garcie-Ferrande et son routier de la mer," *Annales de la Societe Academique de Nantes* vol. 38 (1867): 3–22.

3. E. G. R. Taylor, "A Sixteenth-Century Ms. Navigating Manual in the Society's Library," *The Geographical Journal* 78, no. 4 (1931): 348.

4. E. G. R. Taylor, ed., *A Brief Summe of Geographie by Roger Barlow* (London: Hakluyt Society, 1932).

5. Taylor, *Brief summe*, xv, 181–82; E. G. R. Taylor, "Roger Barlow: A New Chapter in Early Tudor Geography," *The Geographical Journal* 74, no. 2 (1929): 161. Aside from the sections found in navigational manuals, that is, a description of the sphere, the regiment of the North Pole (or the rule of the north star), solar declination tables, and some latitude values, much of the work was representative of a wide-ranging rutter or compilation of basic sailing directions for the oceans and various seas. In addition to his coverage of such coastal areas as Cadiz (where Hercules set his Pillars) or further afield in the Straits of Malacca, Barlow included information from Magellan's voyage and his own account of the exploration of the Rio de la Plata from the failed Cabot expedition. Just as Enciso's work twenty years earlier had "propaganda purposes" by enlarging the valuable spice region in the Spanish sphere of influence to the disadvantage of the Portuguese, Barlow suggested a northern route to the East which remained to be discovered with the benefits of a shorter circuit and sailing always in the daylight.

6. David W. Waters, "English Navigational Books, Charts Globes Printed Down to 1600," *Revista da Universidade de Coimbra* 33 (1985): 240.

7. Gayle K. Brunell, "Images of Empire: Francis I and his Cartographers," in *Princes and Princely Culture: 1450–1650*, vol. 1, ed. Martin Gosman, Alasdair MacDonald, and Arjo Vanderjagt (Leiden: Brill, 2003), 100.

8. Brunelle, "Images of Empire," 99–101.

9. See Helen Wallis, ed., *Jean Rotz, The Maps and Text of the Boke of Idrography* (Oxford: Roxburghe Club, 1981).

10. See David Buisseret, "Monarchs, Ministers, and Maps in France before the Accession of Louis XIV," in *Monarchs, Ministers, and Maps: The Emergence of Cartography as a Tool of Government in Early Modern Europe*, ed. David Buisseret (Chicago: University of Chicago Press, 1992).

11. Peter Barber, "England I: Pageantry, Defense, and Government: Maps at Court to 1550," in *Monarchs, Ministers, and Maps: The Emergence of Cartography as a Tool of Government in Early Modern Europe*, ed. David Buisseret (Chicago: University of Chicago Press, 1992), 59.

12. Barber, "England I," 65; and Waters, "Navigational Books," 240. Waters suggests that, in England, the tide turned with the return of Sebastian Cabot to English service. After Henry VIII's death (1547) Sebastian Cabot (pilot major at the Casa de Contratacion) was invited (or bribed to return) to England to teach navigation. Ten years later, Stephen Borough, brother of

William Borough, was invited to the Casa in Seville. When Boroughs returned to England, he urged Elizabeth to establish a similar institution which prompted Elizabeth to name him Master of Trinity House at Deptford Strand, the recently incorporated guild of mariners.

13. Susan Rose, "Mathematics and the Art of Navigation: The Advance of Scientific Seamanship in Elizabethan England," *Transactions of the Royal Historical Society* 14 (2004): 175.

14. Waters, "Navigational Books," 242–43.

15. See Francis R. Johnson's article, "Thomas Hood's Inaugural Address . . . ," *Journal of the History of Ideas* 3, no. 1 (1942): 94–106.

16. Johnson, "Thomas Hood," 99. "It is not secret Right woorshipfull to any of you all, for we have seene them on our coaste, and heard the thunder of their shot, in how cruell manner the proud disdaynefull insolent Spaniard of late daies hath threatened our Cuntrie, Queene and Citie, with fire and sword, and whatsoever outrageous torment else their villainous mind could worse invent."

17. Johnson, "Thomas Hood," 95.

18. Waters, "Navigational Books," 253.

19. Kenneth R. Andrews, *Trade, Plunder and Settlement* (Cambridge: Cambridge University Press, 1984), 11.

20. Andrews, *Trade, Plunder*, 11.

21. See C. Koeman, "Lucas Janszoon Waghenaer: A Sixteenth-Century Marine Cartographer," *The Geographical Journal* 131, no. 2 (1965): 202–12. Also see Lucas Jansz. Waghenaer, *The Mariners Mirrour London 1588*, trans.Anthony Ashley, facsimile edition (Amsterdam: Theatrum Orbis Terrarum LTD., 1966). The first part included charts and sailing directions for the Western Sea as well as introductory pages forming, in effect, a brief manual of navigation. The second part was for the Northern and Eastern navigation, that is, the North Sea to the Norwegian Sea to the Baltic Sea.

22. Koemen, "Waghenaer," 205.

23. G. R. Crone, "'The Mariners Mirror' 1588," *The Geographical Journal* 119, no. 4 (1953): 455.

24. Crone, "Mirror," 455; Koemen, "Waghenaer," 207; and Rose, "Art of Navigation," 183–84.

25. Koemen, "Waghenaer," 207.

26. Koemen, "Waghenaer," 207.

27. Koemen, "Waghenaer," 208. Also see Lucas Jansz Waghenaer, *Thresoor der Zeevaert Leyden 1592*, facsimile edition (Amsterdam: Theatrum Orbis Terrarum LTD., 1965), with introduction by R. A. Skelton. Skelton writes that that the Spieghel was just too expensive for the regular pilot and the charts within the atlas did not appear to be sold separately. (v)

28. Koemen, "Waghenaer," 208.

29. Koemen, "Waghenaer," 209.

30. Mireille Pastoureau, "Les Atlas Imprimés en France avant 1700," *Imago Mundi* 32 (1980): 56.

31. Buisseret, "Monarchs," 107–20.

32. Pastoureau, "Atlas," 57–59.

33. Pastoureau, "Atlas," 58.

34. C. Passerat, "Étude sur les cartes des côtes de Poitou et de Saintonge antérieures aux levés du XIXe siècle," (PhD diss., Université de Paris, 1910), 67, 76–77; and Pastoureau, "Atlas," 58.

35. Richard Harding, *Seapower and Naval Warfare, 1650-1830* (London: University College, London Press, 1999), 61–62.

36. Harding, *Seapower*, 62.

37. Christine Marie Petto, *When France was King of Cartography: The Patronage and Production of Maps in Early Modern France* (Lanham, MD: Lexington Books, 2007), 57.

38. Stewart Saunders, "Public Administration and the Library of Jean-Baptiste Colbert," *Libraries & Culture* 26, no. 2 (1991): 283–300.

39. Pierre Clément, ed., *Lettres, instructions et mémoires de Colbert*, vol. 3, pt. 1 (Paris: Imprimerie Impériale, 1861-1873), 75, no. 1.

40. Clément, *Lettres*, vol. 3, pt. 1, 464.

41. Clément, *Lettres*, vol. 3, pt. 2, 482. "Il faut penser à faire faire des cartes marines de tous nos voyages, afin de nous tirer une fois de la nécessité de passes par les mains des Hollandois, et de render plus correct ce qu'ils ont fait jusqu'à present."

42. Clément, *Lettres,* vol. 3, pt. 1, 264–65. "Il faut bien prendre garde de tirer de toutes nos navigations, et des journaux qui sont tenus, des connaisssances exactes et fidèles, pour tous ceux qui auront à faire les mesmes voyages, et mesme il faudra s'en servir pour composer des cartes marines."

43. Clément, *Lettres*, vol. 3, pt. 1, 310–311.

44. Clément, *Lettres*, vol. 3, pt. 1, 311, n. 1.

45. Pierre Clément, *Lettres, instructions et mémoires de Colbert*, vol. 5 (Paris: Imprimerie Impériale, 1861–1873), 61. "Vous ne devez pas douter qu'en mesme temps je ne fasse connoistre à Sa Majesté la dépense considerable que ce travail vous a coustée, afin de vous en procurer la recompense. Cependant j'ay donné ordre à celuy qui m'a remis lesdites cartes de les réduire incessament en petites." De Clerville's work (like La Favolliere's work) was among the material Colbert's son, marquis de Seignelay, was to begin compiling.

46. Clément, *Lettres*, vol. 3, pt. 2, 46. "Le Roy veut qu'il soit fait une description exacte de toutes les costes de son royaume, et qu'il soit toujours travaillé dans ses ports à dresser des cartes marines sur les rapports et les journaux de ses vaisseaux de guerre."

47. Olivier Chapuis, *A la mer comme au ciel* (Paris: Presses de l'Université de Paris-Sorbonne, 1999), 101; and Petto, *When France was King*, 58.

48. See Clément, *Lettres*, vol. 5, 176–77, n. 1; *Instruction au Sieur Pène*, 5 February 1678. Also see Clément, *Lettres*, vol. 3, pt. 1, 77–78. "[D]es cartes fort exactes de toutes les sinuosités de la coste, toutes les entrées des rivères, avec les remarques exactes et prises sur les lieux sans s'en fier au rapport de personne, de toutes les rades, hauteurs et bassesses de la mer, dunes, falaises, anses et entrées dans les terres, ensemble de tous les lieux où les ennemis pourroient aborder s'ils estoient assez forts pour faire des descentes ; avec des dessins particuliers de chacun endroit où ils peuvent les faire, et des plans et devis de tous les ouvrages qui pourroient estre faits en chacun de ces lieux, soit pour rompre et gaster entièrement lesdits abordages, soit pour les fortifier en sorte qu'ils puissent estre facilement défendus." (78) See translation in James E. King, *Science and Rationalism in the Government of Louis XIV, 1661–1683* (Baltimore: Johns Hopkins University Press, 1949), 297, n. 197. The government wanted Pène to produce "very exact maps of all sinuosities of the coast, all the entrances of the rivers, with exact notes and taken on the spot without reliance on the reports of anyone, of all the anchorages, highs and lows of tides, dunes, cliffs, creeks and inlets, together with all the places where enemies could land if they were sufficiently strong to make descents; with particular plans of each place where they could make them, and some plans and estimates of all the works which could be made in each of the places, either to break up entirely the said landings, or in order to fortify them in such a way that they may be easily defended."

49. In 1690, Charles Pène, royal hydrographer, was not only to make maps to benefit the navy but also to compile information for the conduct of this work. A memoir of 1690 from Seignelay directs Pène to make charts for the benefit of maritime affairs. He served as the hydrographer of the navy from 1691 to 1701 and at his death in an effort to protect the security of the work he accomplished a seal was placed on his office and all material was to be separated into what involved fortifications and what involved maritime affairs which was moved to the Petits Pères the initial location for the archives of the navy department. See Chapuis, *A la mer comme au ciel*, 159.

50. Chapuis, *A la mer comme au ciel*, 159.

51. Petto, *When France was King*, 58.

52. See J. R. Tanner, ed., *Samuel Pepys's Naval Minutes* (London: Naval Records Society, 1926).

53. See Helen M. Wallis, "Geographie is Better than Divinitie. Maps, Globes, and Geography in the Days of Samuel Pepys," in *The Compleat Plattmaker: Essays on Chart, Map, and Globe Making in England in the Seventeenth and Eighteenth Centuries*, ed. Norman J. W. Thrower (Berkeley: University of California Press, 1978), 1–3.

54. Wallis, "Samuel Pepys," 3. Wallis writes that the Dutch works by Blaeu and Jansson were not readily available and returning "scholars and craftsmen" influenced by new continen-

tal ideas took the opportunity to produce works for the growing class of the middling sort eager to stop relying on foreign sources.

55. Graham Jagger, "Joseph Moxon, F.R.S., and the Royal Society," *Notes and Records of the Royal Society of London* 49, no. 2 (1995): 195. "That for ye desire your Petitioner hath to benefit his Native County with ye most exact and perfect Waggoner in the English Tongue that is yet extant in any Language whatsoever, and that his only feare is that so great and chargeable a work, and so much necessary for our English Seamen, may not when finished answer his pains and cost therein, unless your Majesties Royall approbation should further it."

56. Sarah Tyacke, "Chartmaking in England and Its Context, 1500–1660," in *The History of Cartography, Volume Three, Cartography in the European Renaissance, Part I*, ed. David Woodward (Chicago: University of Chicago Press, 2007), 1746.

57. Tanner, *Pepys*, 135.

58. John Seller, *The English Pilot, First Book* (London, 1671), dedicatory page, BL, Maps, C.22.d.1.

59. John Ogilby, *Britannia* (London, 1675), dedicatory page.

60. Seller, *English Pilot*, dedicatory page.

61. See John Gascoigne, *Science in the Service of the Empire: Joseph Banks, the British State and the Uses of Science in the Age of Revolution* (Cambridge: Cambridge University Press, 1998), 18–19.

62. Tanner, *Pepys*, 317.

63. Johnson, "Thomas Hood," 94–95; and Waters, "Navigational Books," 253. Also see Clements R. Markham, "On the Instruction at Present supplied in this country, in practical astronomy, navigation, route surveying, and mapping," *Proceedings of the Royal Geographical Society and Monthly Record*, 4 (1882): 296–314.

64. Bernard de Neumann, "A Brief History of the Royal Hospital School," in *Mariners: The Website of the Mariners Mailing List*, http://www.mariners-l.co.uk/GreenwichRoyal.html (accessed 17 August 2013).

65. Tanner, *Pepys*, 144–45, my emphasis.

66. Collins surveying work (1681-88) was the first systematic effort to map the coastline of England which prior to this had been piecemeal and/or relied on Dutch works. On 23 June 1681 Charles II specifically commissioned Collins to survey the coasts of Great Britain and was named hydrographer to the king. He was given command of the royal yacht, *Merlin*. See F. H. B. Daniell and F. Bickley, *Calendar of State Papers, Domestic Series of the Reign of Charles II*, volume 22 (London: H. M. Stationary Office, 1921), 328. See Tanner, *Pepys*, 188–89. Samuel Pepys wrote about urging the Trinity House to take it upon themselves to have the soundings of the coast taken especially given that John Adams (topographer), John Seller, and Collins were all seeking support from the government for their survey work. The task fell to Collins being known to the king, Pepys, Naval Officers, and Colonel Legg (Baron of Dartmouth), but Pepys still seems annoyed with the lack of support from Trinity House saying that they were willing to take only two to three of his books once completed. The Trinity House made more formal arrangements by making Collins a "younger brother" of Trinity House, awarding him 50 pounds, but then not allowing him to publish anything without their approval. In March 1682/83 Collins complained to Pepys that he (Collins) had not been given any of the money promised him and furthermore Collins had not received any encouragement from the Admiralty with the exception of the vessel he used to conduct his survey. On the lack of sufficient support for Collins's survey work see A. H. W. Robinson, "Marine Surveying in Britain during the Seventeenth and Eighteenth Centuries," *The Geographical Journal* 123, no. 4 (1957): 450–51; and G. S. Ritchie, "Captain Cook's Influence on Hydrographic Surveying," *Pacific Studies* 1 (1978): 78.

67. Tanner, *Pepys*, 388.

68. Tanner, *Pepys*, 316.

69. Tanner, *Pepys*, 388.

70. The astronomers and engineers working on the Cassini project to map France did considerable work along the French coastline making astronomical observations and setting up a triangulation network. G. S. Ritchie suggests that this technique of marine surveying was only transmitted to England in the early eighteenth century by the Scotsman, Colin Maclaurin, who

travelled to France frequently and struck up a friendship with the more senior Jacques Cassini who was carrying on his father's project to map France. See Ritchie, "Captain Cook," 79–80.

71. Harding, *Seapower*, 149.

72. Harding, *Seapower*, 151. Also see Donald G. Piligrim, "The Colbert-Seignelay Naval Reforms and the Beginning of the War of the League of Augsburg," *French Historical Studies* 9 (1975): 235–62.

73. Harding, *Seapower*, 152.

74. Harding, *Seapower*, 152–53; and Pilgrim, "Naval Reforms," 250–51.

75. Pilgrim, "Naval Reforms," 250.

76. Harding, *Seapower*, 153; and Pilgrim, "Naval Reforms," 251.

77. John A. Lynn, *The Wars of Louis XIV, 1667–1714* (London and New York: Longman Press, 1999), 86.

78. Harding, *Seapower*, 156–61; and Lynn, *Wars of Louis XIV*, 86.

79. Tanner, *Pepys*, 316–17.

80. See Pastoureau, "Atlas," 65. "Pontchartrain dut user de toute son autorité pour obliger les officiers de la marine militaire française à se munir d'un exemplaire de ce recueil; il lui fallut sévir sévèrement et retenir le prix de l'ouvrage sur la solde des officiers."

81. Alice Stroup, *Royal Funding of the Parisian Académie Royales des Sciences during the 1690s* (Philadelphia: The American Philosophical Society, 1987), 54–55.

82. Chapuis, *A la mer comme au ciel*, 103.

83. Tanner, *Pepys*, 388.

84. Tanner, *Pepys*, 324–25.

85. Magentic variation is the variation between magnetic north (which corresponds to the earth's magnetic field) and true north (which corresponds to the meridian of the North Pole). See Alan Cook, "Edmond Halley and the Magnetic Field of the Earth," *Notes and Records of the Royal Society of London* 55, no. 3 (2001): 477–81. Halley compiled several observations of the variation made by pilots and reported to or collected by members of the Royal Society, such as, John Flamsteed, Royal Astronomer, and Peter Perkins, master of mathematics at Christ's Hospital School. After the death of Perkins in 1680, Halley purchased some of his papers which aided his preliminary work for his subsequent voyage and chart.

86. Alan H. Cook, *Edmond Halley: Charting the Heavens and the Seas* (Oxford: Oxford University Press, 1998), 284–85.

87. Cook, *Charting the Heavens*, 285.

88. Edmond Halley, "Advertisement," *Philosophical Transactions of the Royal Society* 22, no. 267 (1700): 728–29. Halley posted an advertisement for navigators, in which he notes that ships entering the Channel (from the West) are falling north of the Scilly Islands and as a result run up the Bristol Channel or Severn Sea with great danger. He writes that the reason for this unplanned action is: "the Change of the Variation of the Compass, and the Latitude of the Lizard and Scilly laid down too far Northerly by near 5 Leagues." (729)

89. Edmond Halley, *A New and Correct Chart of the Channel between England & France* (London: Mount & Page, 1702), BL, Maps *1068 (72). See Cook, "Magnetic Field," 473–90, in particular 477–481.

90. Cook, "Magnetic Field," 484.

91. Cook, *Charting the Heavens*, 285.

92. Edmond Halley, *A New Chart of the British Channel* (London: Mount & Page, 1701), BN, Cartes, GE SH 18 Pf 30 P 11. The chart is given a date of 1701, but is likely 1702 when Mount and Page began publishing it. See date of map at the British Library.

93. Chapuis, *A la mer comme au ciel*, 159–61.

94. Chapuis, *A la mer comme au ciel*, 161.

95. James E. McClellan III and François Regourd, *The Colonial Machine: French Science and Overseas Expansion in the Old Regime* (Turnhout: Brepols, 2011), 61–62.

96. Harding, *Seapower*, 193. A few years later, an attack on the port of Lorient by English ships and their failure to burn out the port left naval action in the Channel at a stalemate. The Battles of Finisterre in the spring and autumn of 1747 that twice captured French battleships coming from Brest once again revealed the weakness of French naval endeavors.

97. Harding, *Seapower*, 199.

98. Jacques Nicolas Bellin, *Nouvelle Carte Reduite de la Manche* (Paris, 1749), BN, Cartes, GE DD 2987 762B.

99. Jacques Nicolas Bellin, *Observations sur la carte de la Manche* (Paris: Chardon, 1749), 1.

100. Bellin, *La Manche*, 1–2.

101. Mary Sponberg Pedley, *The Commerce of Cartography: Making and Marketing Maps in Eighteenth-Century France and England* (Chicago: University of Chicago Press, 2005), 78–79.

102. Petto, *When France was King*, 184-85; and Pedley, *Commerce of Cartography*, 78–79.

103. Papers of the Board of Longitude, *Minutes of the Commissioners*, 16 January 1741/2, RGO 14/5:6, http://cudl.lib.cam.ac.uk/view/MS-RGO-00014-00005/10 (accessed 21 August 2013).

104. William Whiston, *An Exact Trigonometrical Survey of the British Channel* (London: Mount & Page, 1765), BL, Maps *1068 (11).

105. Harding, *Seapower*, 205, 207.

106. Harding, *Seapower*, 160, 207–9.

107. Jacques Nicolas Bellin, *Essai Géographique sur les isles Britanniques* (Paris: Didot, 1757), 3.

108. Bellin, *Essai*, 458–64.

109. Jacques Nicolas Bellin, *Carte Réduite des Isles Britanniques* (Paris, 1757), BN, Cartes, GE SH 18 Pf 17 P 15. In the lower left corner of the map he explained that this one-sheet map was a reduction of the larger sheet map(s), two for England, two for Scotland, and one for Ireland, covering the same area.

110. Jacques Nicolas Bellin, *Partie Meridionale de l'Angleterre* (Paris, 1757), BN, Cartes, GE DD 2987 2023B.

111. Bellin, *Essai*, 291–94.

112. Bellin, *Essai*, 293–94.

113. Bellin, *Essai*, 437–38.

114. "A Description of the Isles of Jersey, Guernsey, Alderney, and Sark: With a Correct Map," *London Magazine* 22 (1753): 543.

115. In the Archives Nationale there are several documents regarding the possibility of an invasion of Britain. AN MAR B/4/74, f. 2 (1756) includes an analysis by sieur Vivant de Maissagues regarding a "descente" on England. AN MAR B/4/82, f. 115, Bellin's *Remarques* on the Channel Islands. AN MAR B/4/82, f. 135 (1756) includes instruction for Duc d'Aiguillon and comte de Raymond for an enterprise against the island of Jersey and other documents such as M. Silhouette's comment on the project to descend on Britain and a description of the SE coast of England ordered by M. Pontchartrain and the usefulness of Jersey in such a descent.

116. Bellin, *Essai*, 435. Maps of the Channel Islands produced at the Dépôt and by the order of Machault: *Carte Reduite des Isles de Jersey, Grenesey et d'Aurigny avec les costs de Normandie et de Bretagne qui en sont voisines* (Paris, 1757), BN, Cartes, GE DD 2987, 1098; *Carte de l'isle de Grenesey* (Paris, 1757), BN, Cartes, GE DD 2987, 1109, on which Bellin noted in the cartouche that this map was drawn by an English manuscript recently acquired for the Dépôt; *Carte réduite de l'isle Aurigni* (Paris, 1757), BN, Cartes, GE DD 2987, 1112, by Bellin on an English manuscript; *Carte de l'isle de Jersey* (Paris, 1757), BN, Cartes, GE DD 2987, 1104, drawn from an English map by Cpt. Lempriere in London 1755.

117. J. B. Harley, "The Bankruptcy of Thomas Jefferys: An Episode in the Economic History of Eighteenth Century Map-Making," *Imago Mundi* 20 (1966): 31.

118. Harding, *Seapower*, 211.

119. Thomas Jefferys, *Seat of War on the Coast of France* (London, 1758), BL, Maps, K. Mar. III.9.

120. "Historical Chronicle," *Gentleman's Magazine* 28 (1758): 288.

121. "Action at St. Cas Bay," 435-36; "Reflexions on our late attempts in France," 495–96; "Account of the late expedition to the coast of France," 516–21, *London Magazine* 27 (1758).

122. "Reflexions," 496.

123. "Reflexions," 496. "I should be glad to know, for what reason we remained in France, after the design upon St. Maloes was laid aside; with what view we penetrated so many days march into the country; neglected the repeated intelligence we received; communicated, by beat of drum, our midnight motions, to an enemy of double our force; loitered near seven hours in a march of three miles; and, lastly, began the re-embarkation of the troops, at a place where no proper measures had been taken for their cover and defence."

124. *London Magazine* 27 (1758), 519–20. A pamphlet, *An Impartial Narrative of the last Expedition to the Coast of France*, reflected on the leadership. "[I]n choosing a place (St. Lunaire Bay) for landing troops, he 'departed very much from his usual prudence and good conduct, in landing the troops in unsettled weather, on a beach, from whence, except the wind blew directly off shore, and that but slightly, he could not reimbark them.'" (519) Being unable to re-embark, troops were orderd to St. Cast where the battle ensued. "No ground could have been more advantageous for the enemy." (520)

125. *Gentleman's Magazine* 28 (1758), 588, 587–90.

126. Harding, *Seapower*, 205.

127. James Pritchard, *Louis XV's Navy 1748–1762: A Study of Organization and Administration* (Montreal: McGill-Queen's University Press, 1987), 15.

128. Harding, *Seapower*, 214.

129. Harding, *Seapower*, 214.

130. Harding, *Seapower*, 215.

131. *Gentleman's Magazine* 29 (1759): 576, my emphasis.

132. *London Magazine* 28 (1759): 640.

133. Thomas Jefferys, *A Description of the Maritime Parts of France* (London, 1761), dedication.

134. *The Critical Review: or, Annals of Literature* 11 (1761): 205–7.

135. *The Critical Review: or, Annals of Literature* 11 (1761), 206.

136. *The Critical Review: or, Annals of Literature* 11 (1761), 207.

137. *The Critical Review: or, Annals of Literature* 11 (1761), 208.

138. *The Critical Review: or, Annals of Literature* 11 (1761), 208.

139. Thomas Jefferys, *Le Petit Neptune François: or the French Coasting Pilot* (London, 1761), iv.

140. *The Critical Review: or, Annals of Literature* 11 (1761), 209.

141. Jacques Nicolas Bellin, *Carte reduite de La Manche* (Paris, 1763), BN, Cartes, GE SH 18 PF 30 P18.

142. Harding, *Seapower*, 232.

143. *Archives des Affaires Etrangeres, Paris* 52 (1772), f. 180–224, 30 July 1772, Mémoires "sur l'Angleterre," avec étude de projet de descente par la Tamise, Portsmouth ou Plymouth, par le comte DE GUINES.

144. *Archives des Affaires Etrangeres, Paris* 54 (1775), f. 313.

145. Chapuis, *A la mer comme au ciel*, 190.

146. Chapuis, *A la mer comme au ciel*, 190. Translation of French found in Pedley, *Commerce of Cartography*, 193.

147. McClellan and Regourd, *Colonial Machine*, 61–62, 190.

148. Chapuis, *A la mer comme au ciel*, 204. After having been denied permission to succeed to the chair of hydrography at Le Havre in 1766 over concerns of his ability, Degaulle was vindicated as he was entered into the Académie de Rouen in 1774 on account of his contributions to navigational instrumentation.

149. Jean-Baptiste Degaulle, *Nouvelle carte réduite de la Manche de Bretagne* (Havre, 1773), BN, Cartes, GE AF Pf 35 (73).

150. Jean-Baptiste Degaulle, *Nouvelle carte réduite de la Manche de Bretagne* (Havre, 1778), BN, Cartes, SH 18 Pf 30 28/2

151. Chapuis, *A la mer au comme ciel*, 204.

152. Harding, *Seapower*, 243–44.

153. Susanna Fisher, *The Makers of the Blueback Charts: A History of Imray Laurie Norie & Wilson Ltd* (Ithaca: Regatta Press, 2001), 19–21.

154. Bennett was no longer a partner in the business by 1785 having suffered a period of insanity beginning in 1781, confined for 9 months in 1783, and brought to court by Sayer in 1784 in order to dissolve the partnership. See Fisher, *Blueback Charts*, 57.
155. Fisher, *Blueback Charts*, 56.
156. Fisher, *Blueback Charts*, 58.
157. Pedley, *Commerce of Cartography*, 194; and Chapuis, *A la mer comme au ciel*, 190.
158. Fisher, *Blueback Charts*, 58.
159. Arthur Underhill, *A Summary of the Law of Torts*, 5th edition (London: Butterworths, 1889), 296.
160. Fisher, *Blueback Charts*, 7.
161. Pedley, *Commerce of Cartography*, 195; and Chapuis, *A la mer comme au ciel*, 202–3.
162. Pedley, *Commerce of Cartography*, 195.
163. Pedley, *Commerce of Cartography*, 194–95.
164. Fisher, *Blueback Charts*, 19–20.
165. John Hamilton Moore, *Chart of the British Channel* (London, 1786), BL Maps *1068(58).
166. Fisher, *Blueback Charts*, 21.
167. N. R. Essenhigh, Rear Admiral, "The Hydrographic Office: The Next 200 Years," *Naval Review* 83, no. 2 (1995): 90.
168. Essenhigh, "Hydrographic Office," 90.
169. Fisher, *Blueback Charts*, 10–11.

Chapter Four

Paper Encroachments

Colonial Mapping Disputes in the Americas

> [M]aps are from the Nature of them a very slight Evidence, Geographers often lay them down upon incorrect Surveys, copying the Mistakes of one another; and if the Surveys be correct, the Maps taken from them, tho' they may shew true Position of a Country . . . can never determine the Limits of a Territory. —*Memorial of the English Commissaries*, 11 January 1751

In the late seventeenth century, as England and France had each just come to produce their national maritime atlases describing their coasts, home waters, and the Channel, they regarded the state of their colonial possessions in North America. After the cessation of the Nine Years' War, its American counterpart—King William's War—and the signing of the Treaty of Ryswick (1697), English officials at home and in the colonies raised their voices in indignation over French pretensions in North America and their encroachment on English colonial regions. To the north, the French were expanding the territory of Nova Scotia to the Kennebec River, and along the *back* of English territories from New England to the Carolinas, the French traveled along the St. Lawrence River to the Great Lakes southward and westward extending "their pretensions to the propriety of all the countries bordering upon the said river and lakes."[1] English patents granted lands "from sea to sea" without clear land boundaries, and the Council of Trade and Plantations writing to William III's Secretary of State, James Vernon warned, "it will turn to the very great prejudice of England, and most particularly of New York and other[s] [of] His Majesty's provinces in America, whose frontiers are furthest extended towards the places so claimed by the French."[2] In addition to alarm over French encroachment on English lands, there was great concern that the French would continue to undermine English trading in the area, sour relations between the English and their native allies, and agitate France's native allies to war.

The English were not alone in their cries against foreign encroachment, as the French had denounced English intrusions on their territory. The French ambassador asked that orders "be sent both to New York and to Boston forbidding such encroachment. The King of France is prepared to agree to a delimitation of boundaries to prevent the recurrence of such incidents."[3] The governor of New York, Thomas Dongan, had written months earlier of the concerns over cross-regional trading and provided a map to the king of England so he might view "how the land lies."[4] Disputes over boundaries were only just beginning, as in the aftermath of the Nine Years' War (1688–1697), France sought to expand its colonial holdings in the Americas.

In the eighteenth century, in an effort to extend the influence and power of its early modern state, Britain and France moved well beyond their own interiors to the Americas, and their map makers were prepared to serve their state with well-reasoned documents touting a scientific authority on their maps, in their *mémoires* (reports), in their correspondence, in the periodical literature, and with their methodology. They were ready to serve national interests often in response to what they perceived as "paper encroachments" by one another. Many decades prior to Britain, however, France had begun to create and support "an elaborate scientific and technical infrastructure that was not merely tapped on occasion to aid colonization, but which quickly became integral to the process, the Colonial Machine."[5] As a consequence, Paris had become the center of cartographic activity by the second half of the seventeenth century, and of hydrographic activity by the second quarter of the eighteenth century, and among its many map makers, Guillaume Delisle, Jacques Nicolas Bellin, and Jean Baptiste Bourguignon d'Anville benefited from their connections to this infrastructure and institutions of power such as the court, the Académie, and the naval ministry as well as knowledgeable correspondents and voyagers who shared their information all of which placed these map and chart makers at the very center of the associated cartographic (or hydrographic) activity.

While London map makers such as Herman Moll, Henry Popple, John Green, and Thomas Jefferys similarly enjoyed connections to the court, the Royal Society, the Board of Trade and Plantations, and the Admiralty, they did not have the benefit of an infrastructure with "the level of organized science and state-sponsored expertise oriented toward the colonies."[6] They were, however, well aware of the importance of the encouragement of the French state for geographic activity; near mid-century, John Green acknowledged this beneficial relationship in France. "The French have long engrossed the Care of Geography to themselves, and it must be confessed have brought it to greater perfection than any Nation in Europe. This is owing to the proper Encouragement which its Professors (never so numerous as at this

Time) meet with in France: a Country so fruitful of Patrons for the Sciences."[7] Nonetheless they did not intend to allow French geographers to use or misuse information to the benefit of the French state. At that point prior to the start of the Seven Years' War, in 1753, Green argued that the English were just as capable as the French to improve the Sciences, but the proper encouragement was necessary to ensure that this important work did not fall into the hands of say, engravers, who may not have the same motivations to produce high quality work.

Green's comments highlight elements for a comparative study, that is, the importance of state support for the work of geographers and the role of the map printing trade. These two points coupled with the socially constructed nature of maps and charts as well as the paradoxical rhetoric of the boundary disputes in which both sides decried the unreliability of maps and yet sought them out (even supported their creation) are the topic in this chapter on English and French maps and charts of North America. Nowhere is the point made by John Pickles regarding map making more clear than in the colonial enterprise: "Maps and mappings precede the territory they 'represent.' . . . Territories are produced by the overlaying of inscriptions we call mappings."[8] While the English may have argued that maps were useful for locating the position of a territory, but not its limits, they proceeded as did the French to support the production of maps that defined the limits of territory. Maps were powerful tools for these states as they made claims to lands and set boundaries in a politically volatile climate. Maps embody a dual nature: the plain, factual map serves a referential function, that is, location of river, city, or town, at the same time that it serves a discursive function, that is, here, a rhetoric of power through possession but touting that power as born of a scientific methodology or superior information.[9] Moreover, the presentation in this chapter highlights this tension between the representational (or referential) aspects of these maps, such as denoting boundaries, cities, rivers, and the like, and the discursive nature of them, that is, rhetoric of power displayed *on* the maps and in accompanying promotional reports. In addition, the claims of applying a scientific methodology to the cartographic practice served both to give greater credence to the representation as well as to provide greater authoritative power to the discourse.

Much of the mapping for North America embraced an effort to have cartographical knowledge of place(s) but also a means of declaring territorial legitimacy in the face of English or French encroachment, depending on one's point of view. This chapter focuses on two examples from the cartographic debates in the colonies. First, Guillaume Delisle's map, *Louisiane* (1718), caused quite a stir as Britain perceived French territorial transgressions against their well-established colonial possessions. The response included not only audible complaints from the Board of Trade and Plantations [hereafter, Board], but also published English maps. Second, several decades later, Britain and France found themselves once again at odds over the territory of Acadia, and, despite a wariness on both sides concerning the use of maps to solve the dispute, their mémoires included a discussion of evidentiary maps. While there was no official map of settlement, English and French cartographers would produce maps and accompanying reports allied with their national position. An English audience would also see these debates played out in the pages of popular periodicals such as the *Gentleman's Magazine* or the *London Magazine* giving voice to the indignant literate class against the "paper encroachment" or pretensions of the perfidious French state.

GUILLAUME DELISLE'S LOUISIANE (1718)

Delisle devoted his greatest geographical attention to the mapping of the Americas.[10] On his first map of North America (1700), he not only noted that his work was drawn from the observations of members of the Académie and most recent reports, but also that his maps were created by a unique methodology.[11] Delisle included an *avertissement* below the cartouche of the scale of measures in which he acknowledged that his maps were different than those already published (i.e., by those numerous royal geographers), but that his changes were not arrived at by inadvertence but born out of a more thoughtful, intentional, or deliberate approach. In other words his work reflected the rational and critical ideology of the Enlightenment, that is, a methodology based on empiricism, and the rational and systematic analysis of textual sources. He directed readers to the work, *Nouvelle Introduction à la Geographie* first published by Nicolas Sanson, who had passed on the method of textual analysis (the exegetical method) to Guillaume's father, Claude Delisle, an historian and geographer, tutored by Sanson. With his father's influence, Delisle entered the Parisian map trade not only with an unsurpassed talent, but also the objective to continually update maps based on a reasoned utilization of observations, measurements, and textual sources, rather than merely imitating the works that came before, making a few changes, and publishing supposedly new maps for commercial benefit.

A pupil of the famous Cassini and in 1702 a member of the Académie, Delisle—with his father as collaborator until 1718—produced for the court maps of America containing the most recent information.[12] The Delisles were able to capitalize on the circumstances because of the dearth of geographical knowledge of the New World and because of France's growing interest in establishing colonies and trading centers there, especially in Louisiana. Delisle's reputation and geographical works would come to force English map makers to produce works in opposition (though often using Delisle's work as a source) to prevent "paper encroachments."

Among the sources for the Delisles on the Americas were the maps and reports of Jean Deshayes (d. 1706), royal hydrographer in New France, and Jean-Baptiste Louis de Franquelin (1651–c. 1712), cartographer of New France.[13] For the lands further south, the Delisles had the information from René Robert Cavelier, Sieur de La Salle (1643–1687), who voyaged further than Louis Jolliet (1637–1675) and Jacques Marquette (1646–1700), reached the Mississippi delta in the Gulf of Mexico, and in 1682 took possession of the river and the land, which he called Louisiana, in the name of the king, Louis XIV.[14] La Salle's 1684 expedition also marked the direct involvement of the Delisles in obtaining information on the Americas. The elder Delisle, in a letter to Cassini, reported that he communicated with members of this expedition to the Gulf of Mexico both prior to their voyage and after their return.[15] In the succeeding two decades, after the murder of La Salle in Texas (1687), the Delisles obtained more information on Louisiana from the reports of Pierre-Charles Le Sueur (c. 1657–1704) and from the two expeditions of his cousin, Pierre Le Moyne, Sieur d'Iberville (1661–1706), the first of which established the settlement of Biloxi. On his second expedition (1699–1700), the court instructed him to gather information on the feasibility of establishing a colony along the Mississippi. Similarly, the elder Delisle, for the Delisles' map making efforts, gave d'Iberville a map, a report, and clear instructions of what details to gather in order to improve the geographical knowledge of the area.[16]

In 1697, Louis Hennepin (1626–1705), a Franciscan missionary and explorer who had accompanied La Salle on his second expedition into the western regions of New France and south to the Mississippi River, published in Utrecht his *Nouvelle Découverte*. Having fallen out of favor with the French crown, Hennepin (in search of new patronage) directed the dedication of this new work to William III. He urged him to bring the "light of Christianity" to the people of America and at the same time benefit the public by the great advantage of commerce with America.[17] He accompanied the book with a map also dedicated to William as a graphic aid to advocate for further English settlement in these regions. While there was already correspondence warning of French encroachments and urging vigilance on the part of the officials, English maps of Canada and the region of the Mississippi River would remain derivative of French works until the second half of the eighteenth century.[18]

About a decade later, on 10 July 1709, Thomas Nairne (d. 1715), an Indian agent, included a map in his report to the secretary of state, Charles Spencer (1675–1722), Earl of Sunderland, on which he hoped to show what lands the English were in possession of, what threat the French at nearby Mobile were, and that South Carolina was of such strategic importance that it should not be neglected.[19] The French at Mobile were well situated for Indian trade as well as to stir up animosity amongst the natives towards the

English. Nairne travelled widely in the area and advised that Britain expand its presence and attention to trade with the Indians not only to benefit general colonial trade but to keep the French in check. Moreover, he suggested that a study of all the maps and descriptions of the area would support the position that the English consider establishing another colony to the west of the mouth of the Mississippi in order to take advantage of strained relations between the Indians and the Spanish and develop trading relations with both. The challenge remained, however, of keeping the French from encroaching on South Carolina. Nairne's manuscript map was published in an abridged version as an inset on Edward Crisp's 1711 map of the province of Carolina.

Even after the Treaty of Utrecht (1713), the English remained concerned over French pretensions along the strategic axis extending from New France to the Gulf of Mexico. In a March 1715 report to Alured Popple (d. 1747), secretary of the Board, Richard Harris (colonial trader) continued to express English concerns over French pretensions.

> [T]he great schemes formed by France for founding a universall power in America as well as in Europe may one day take place she hath extended her acquisitions from Quebeck on the back of New England very far up the great Lake of Canada almost as high as the back of Carolina into the very heart of the great Continent of North America to the heads of some rivers which 'tis said have a communication with the great River Misissippi [sic] which falls into the Gulf of Mexico at the mouth whereof the French have made a great settlement.[20]

As yet there were no surveys by triangulation by either English or French surveyors, but the English with numerous settlements east of the Appalachians had surveyed particularly strategic and vulnerable areas such as the Carolinas and had with difficulty explored the region around the Mississippi River. Nairne had utilized the Mississippi River as a boundary between the English and French possessions. Delisle's maps of America indicated that the French had other ideas.

Using his manuscript map on the course of the Mississippi River drawn from the information from the travels of d'Iberville and Le Sueur, in 1703 Delisle published two maps covering the geographical area of North America—Canada and the area of the Mississippi River—and presented them to the minister of the navy, Jérôme Phélypeaux, comte de Pontchartrain (1674–1747). Capitalizing on the court's desire to see the promise of French influence in the New World, these maps, utilizing observations and printed and manuscript reports, provided the minister with the most recent information and specifically described the Mississippi River, an area of great interest to the French crown and most certainly not the border for South Carolina.[21]

Delisle continued to map the area as more information came to light. His correspondence with Père Jean Bobé, a Lazarist priest of the Congregation of the Mission, a chaplain at Versailles, and himself an author of several works on the area of Louisiana, yielded him important contacts and direct information for his New World maps.[22] Bobé, in effect, acted as a broker of information or patronage for Delisle, connecting him further with members of the inner circle surrounding Pontchartrain, including Antoine-Denis Raudot (1679–1737), the son of the intendant of New France and one of the future directors of the Compagnie des Indes. In a letter carrying the supposed date of 15 January 1710, Bobé wrote to Delisle that Raudot was "in favor" of him and that if he (Delisle) sent his world map, Bobé would present it to Raudot.[23] Bobé also reproved Delisle for not putting the line separating Louisiana from Mexico further west in order to give more territory to France; he suggested that Delisle use the land of the Cenis—taken possession of in the name of Louis XIV by La Salle in 1686—as the boundary between Louisiana and Mexico.[24]

In a later letter, Bobé wrote that Raudot was impressed with Delisle's world maps (1714), but there was concern over the representation of boundary lines. "Mr Raudot begs you to remove from your plate the *dots that you have put in to mark the limits of Louisiana, California, New Mexico, etc. the court does not agree to the limits assigned by geographers, yet foreign nations use our maps against us when we discuss important questions with them*."[25] In 1712, the crown officially revived its "company project" in Louisiana with the letters patent awarding the wealthy financier, Antoine Crozat (1650–1738), a fifteen-year monopoly of the commerce of Louisiana.[26] Throughout the negotiations for the Treaty of Utrecht (1713), both the Dutch and the English expressed their alarm and displeasure at Crozat's monopoly, which they viewed as a further insult given the "Indies profit" France would benefit from by its union with Spain. Such disapproval was expected from international competitors, but even within Louis' council there was great trepidation at turning the colonial commerce over to a monopoly, the Company of Louisiana.[27] Raudot's concerns represented a government which was not yet willing to go public with boundary lines that pushed France's colonial influence further into the lands of other countries. In effect, since maps *produce* the territory, French officials had concerns regarding what the map would define. At the beginning of 1715, France, which recently had completed treaty negotiations with the rest of Europe to bring an end to the War of Spanish Succession, was not in a position to start another international conflict.

By the end of the same year, however, Soupart, a French officer and pilot, completed a survey of the area of the Gulf Coast and the mouth of the Mississippi River.[28] Shortly after, Père François Le Maire, a secular priest sent on mission by the *Séminaire des Missions Étrangères* and one of Delisle's primary sources, sent four maps to France based on Soupart's survey.[29] In 1718, with peace treaties signed, new information to support French claims in the area (Soupart's survey), and a new government under the Regency, Delisle published his map of Louisiana with territorial limits that led to a volatile cartographic dispute with England and Spain. Delisle's work, *Carte de la Louisiane et du cours du Mississipi*, became the mother map (or source map) for all succeeding maps of the Mississippi because of its accurate depiction of the lower Mississippi River and the surrounding areas. He based his map of Louisiana (1718) on the most recent information from Soupart's survey and Le Maire's works.[30] And in an effort to support the territorial claims and protect French trading interests, his work reflected the French custom that "the entire watershed of a traveled waterway was encompassed by the right of 'discovery.'"[31] Furthermore, with the advent of scientific knowledge, that is, Soupart's survey, the referential function of the map was more secure and no longer in conflict with the discursive function. France needed maps of the Gulf Coast area and the course of the Mississippi River if it were to capitalize commercially and militarily along the trading axis (Saint Lawrence-Great Lakes-Mississippi) and along the strategic axis (Quebec-Great Lakes-Mobile, the site of d'Iberville's second settlement).[32] In a country whose boundaries were yet to be clarified, maps were powerful political tools in the international arena.

THE ENGLISH RESPONSE

Internationally, Delisle's map of Louisiana generated a boundary dispute that lasted for several decades into the Seven Years' War (1756–1763). In the map, Delisle extended the areas under French control in direct opposition to English claims by pushing the English colonial border further east than the Appalachian frontier and then further aggravated them by claiming that Carolina was named for the French king Charles IX (d. 1574) and taken possession of in the sixteenth century. In late November 1720 the Governor of New York, William Burnet (1688–1729), wrote to the Commissioners of the Board decrying the impertinence of the French in setting up a trading house near the "fall of Niagara" and trading with the Indians to the detriment of British.[33] The land had been "granted" by the Sennekees to Britain, and as a strategic axis of communication "between Canada and Micessipi [sic] by the River Ohio," the French could not be allowed to remain there uncontested. "[T]he use the French made of our easiness to them," Burnet suggested, emboldened them in their pretensions, and Britain had to act to break "the neck of this trade."[34] "This is in open defiance of the King's right, the Sennekees having granted that land to the Crown of Great Britain before the French had ever been there."[35] Burnet wrote that even more evidence of French perfidy was found in Delisle's recent map, graphic evidence of their "paper encroachments."

> I observe in the last mapp published at Paris with *privilege du Roy par Mr. de Lisle* in 1718 of Louisiana and part of Canada, that they are making new encroachments on the King['s] territories, from what they pretended to in a former map, published by the same author in 1703 of North America, particularly all Carolina is in this new map taken into the french [sic] country, and in words there said to belong to them, and about 50 leagues all along the edge of Pensilvania and this province taken into Canada more than ever was in their former map.[36]

In the years leading up to (and soon after) the publication of Delisle's map, governors and other officials from Boston to Charleston wrote to the Board concerning French encroachments which put English colonial commerce and the safety of its plantations in jeopardy. For example, Richard Beresford, a Charleston merchant and member of the governor's council, wrote to the Board in late 1717.

> It has been long foreseen and complain'd of, that ye French had a design to dispossess us of all our Plantations in the Continent of North America towards which they have been very industrious to make discoverys in those parts. . . . It is therefore humbly propos'd to ye Government that they wou'd be pleas'd to think of proper methods to assert the sovereignty and honour of ye Crown of England against such encroachments.[37]

The English position had clearly been undermined by a lack of attention and geographical information. Long before John Green commented on the encouragement enjoyed by French geographers, the British government not only knew that Paris was the center for the map trade, but also that the best maps (of America) could be found there. In July 1715 the Board wrote to Secretary Stanhope to instruct the British minister at the French court to have someone purchase and collect "the best maps of America."

> It being necessary for us on several occasions, to consult ye maps of the Plantations, in order to our having not only a true idea of those parts, but also of the matters before us, we have made all the inquiry possible for such maps, and can find but few here that are to be depended on; And therefore we think if for H.M. service, that H.M. Minister at the Court of France, be desir'd to direct some person there, to make a collection of the best maps of America, that can be had, whether general or particular maps of Islands, Provinces and Settlements made by any Europeans there.[38]

Perhaps in part to address the lack of information as well as the latest French geographical production, that is, Delisle's map, Whitehall directed the Board to create a circular letter to the Governors of the Plantations on the Continent of America.

> It being necessary for H.M. Service and for the benefit of the Plantations, that the limits or boundaries of the British Colonies on the Continent of America, should be distinctly known and marked out, more particularly so far as they may border on the settlements made by the French or any Foreign Nations, we desire you to send us, as soon as possibly can, the best informations you can get upon your Government, together with a chart or map, and the best accounts and vouchers you can obtain to support the same, more particularly with relation to any settlements that may have been made by the English on the frontiers towards the Lakes and mountains: You will at the same time inform us, whether the subjects of any foreign Prince have made encroachment on the Province/Colony under your Government.[39]

A year later the Board writing to Lord (John) Carteret (1690–1763) urged the government to encourage the expansion and settlement of Her Majesty's plantations more towards the Appalachian Mountains as an effective means of thwarting French pretensions. In the same summer, Governor Burnet writing from New York to Lord (Charles) Townshend (1674–1738) on French efforts in the region expressed what had become quite evident, "[w]hile the limits are unsettled the French loose [sic] no time to make all possible encroachments."[40] Such a situation begged for map!

The initial reply from the English came in the form of a map by Herman Moll (c. 1654–1732), a contemporary of Delisle who denounced the French territorial claims in his maps of North America.[41] Moll, an engraver, mapmaker, publisher, and consummate self-promoter, peppered his maps and atlases of America with remarks promoting British interests and stern warnings against French encroachment. For example, in his 1720 map of North America, Moll depicted the boundaries claimed by France as well as those "they allow ye English."

> The French Divisions are inserted on purpose, that those Noblemen, Gentlemen, Merchants &c. who are interested in our Plantations in those Parts, may observe whether they agree with their Properties, or do not justly deserve y Name of Incroachments; and this is ye more to be observed, because they do thereby Comprehen within their Limits ye Charakeys and Iroquois, by much ye most powerfull of all ye Neighbouring Indian Nations, the old Friends and Allies of the English, who ever esteemed them to be the Bulwark and Security of all their Plantations in North America.[42]

In response to French ludicrous claims in the Carolinas, Moll provided a short history lesson (on the map) of English arrival and settlement, that is, the grant by the English king, Charles II, to the Lord Proprietors in 1663. He even used the "original draughts" of Beresford, the stalwart opponent of French pretensions from the South Carolina Assembly.

Moll, a veteran of the South Sea Bubble affair, was not a newcomer to colonial promotion. At the famous Tory coffeehouse, Jonathan's, he entertained with his friends—the scientist, Robert Hooke (1635–1703), the writer, Daniel Defoe (1660–1731), and the pilot and English buccaneer, William Dampier (1651–1715)—a common vision of English expansion.[43] Moll's 1720 map dedicated to Thomas Bromsall, Esq., was a graphic document to support not only English territorial rights to the lucrative fishing grounds and fur trade along the Niagara River, Newfoundland, Canada, and Hudson's Bay, but also the security of the English settlements whose western extent was being encroached upon by the French.

In the upper left corner of the map was a view of Fort Susquehanna, initially occupied by the Susquehanna, allies of the French. Built with French assistance at the headwaters of the Susquehanna River, the fort was representative, according to the English, of another artifice utilized by the French to encroach.

> [A]s the French are convinced that a charter without possession can never be allowed by the laws of the nations to change the property of the soil, they have employed another artifice, and without embarrassing themselves about former discoveries made by the subjects of other princes, have built small forts at the heads of lakes and rivers along the vast tract of land from the entry of the river of St. Lawrence to the mouth of the Mississippi into the Bay of Mexico not with the intention to bound their territories but to secure what they already have until a more favorable time gives them the opportunity to intrude more on your majesty's land.[44]

The fort, having been seized by the Iroquois, under English protection, was now a symbol of English land possession.

Complaints of French encroachment on trade with the Indians as well as on English territory continued over the course of the next three decades. New York officials were particularly incensed at French actions, and Governor Burnet ordered a map to counter French claims. The 1724 small map published only in New York, *A Map of the Countrey of the Five Nations belonging to the Province of New York and of the Lakes near which the Nations of far Indians live with part of Canada*, was merely a section of Delisle's map over which officials inserted "their version of possession."[45]

In 1734 the French constructed Fort Saint Frédéric at Crown Point, New York, on Lake Champlain and had designs on Tierondequat (i.e., Irondequoit) on Lake Ontario. Lieutenant-Governor George Clarke (1676–1760) wrote to the Thomas Pelham Holles, Duke of Newcastle (1693–1768), and the Board including an updated version of the 1724 map noting the location of Fort Frédéric and Tierondequat in red. "The French pretend to claim all the lands so far as the springheads of any rivers or waters that empty themselves into any of the lakes that disembogue into the river of St Lawrence."[46] Over the course of the first half of the eighteenth century, New York officials—who were not alone in their encroachment concerns—felt a constant need to put their plight for protection/attention before the eyes of the government in London. It was in response to such claims that the Board had approved the work of Henry Popple (d. 1743).

In 1733 Henry Popple published *A Map of the British Empire in America with the French and Spanish Settlements adjacent thereto* which was to be a "practical tool for political ends," that is, public legitimacy of British territorial rights.[47] On the map an engraved text indicated that Popple had accomplished the map with the "approbation" of the Lord Commissioners of the Board. Given the circular letter from 1721 sent to all the governors of the British plantations, it also appears that Popple's map was the first fruits of the resulting reports: "great care has been taken by comparing all the maps, charts, and observations that could be found, especially the authentick records and actual surveys transmitted to their Lordships by ye governors of the British Plantations and others to correct ye many errors committed in former maps." In addition, Popple's map, dedicated to the Queen, was evaluated by the well-respected astronomer, mathematician, map (and chart) maker, and member of the Royal Society, Edmond Halley whose report was also engraved on the map. Halley found the map to be laid down with great accuracy and to depict the possessions of that part of the globe "more truly than yet extant."[48]

Popple's map, however, received a mixed reception for having used French sources, the very root of the dispute in the first place. Rather than asserting strong British territorial rights in the region of the Great Lakes, he left much of the lake region in the hands of the French.[49] Popple had at his disposal the manuscript map by Colonel "Tuscarora Jack" Barnwell (1671–1724), "one of the most important maps of colonial southeastern North America," but with which he, again, did not assert strong British territorial claims in the expansion of South Carolina.[50] The Commissioners distanced themselves from Popple's work, and while they admitted approving the undertaking of a new map, they denied giving approval for the map's execution, that is, engraving and publication.[51] English maps of the colonies would remain primarily derivative of French works for a few more decades.

In early 1738, Cadwallader Colden (1688–1776), surveyor-general for New York, strongly expressed a similar view. He sent to Governor Clarke, responses to queries from the Board on the governance and security of the province. In describing the great advantages for trade in the region due to the numerous lakes and rivers, Colden acknowledged: "These advantages I am sensible cannot be sufficiently understood without a map of North America. The best which I have seen is Mr. De L'Isle's map of Louisiana published in French in 1718."[52] The best map available then to evaluate the possible expansion of commerce was a work that shrunk the very territory over which the British—or at least the occupants of the province of New York—intended to expand their trade. Furthermore, he wrote: "If we are to judge of the pretensions of the French by the maps lately published in France by public authority they not only claim this part of the country and the countries of the Five Nations depending on New York but likewise a considerable part of what is actually settled by the inhabitants of New York. The English maps are such servile copies of the French that they mark out the boundaries between the English and French with the same disadvantage to the English that the French do."[53]

For example, in 1736 d'Anville published a map of North America for his patron, duc de Chartres, on which the western boundaries of the province appeared far more beneficial to French possession than English presence (or influence) would have revealed.[54] Further to the south, this same map is silent on the establishment of the English province of Georgia. The 1732 charter for this "adventure in social engineering" established the colony "between the Savannah and Altamaha rivers and from the heads of those rivers 'westward to the South Seas' [i.e., Pacific Ocean]."[55] James Edward Oglethorpe (1696–1785), member of the House of Commons and architect of the group, Trustees for the Establishment of the Colony of Georgia in America, worked tirelessly to help promote the success of the colony. While Parliament would not support the emigration of skilled craftsmen, there was less opposition to providing an opportunity to England's unemployed or Europe's Protestant refugees, but certainly in addition to the project's mission as a "social safety valve," its location just below South Carolina meant it was to be a buffer zone between English possessions and French and Spanish territories. Spanish concerns aside, Georgia was to act as a hindrance to French encroachment from nearby Mobile, Biloxi, or New Orleans. As a English colony, Georgia also needed to fill the mercantilist needs of the mother country. With Benjamin Martyn (d. 1763), the secretary for the trustees, Oglethorpe produced a map to help promote this Georgia scheme.

Louis De Vorsey suggests that the map that accompanied the promotional pamphlet, *Reasons for Establishing the Colony of Georgia* (London, 1733), "was judiciously compiled to present Georgia in a geopolitically favourable and ostensibly secure setting."[56] While the position of the name of the colony, Georgia, was relatively confined between the Savannah and Altamaha rivers in contrast to an earlier state of the map that spread the name out virtually from the Mississippi River to the Atlantic Ocean, the map was surprisingly reserved in its depiction of the presence of both French and Spanish interests.[57] The French are dispatched to the west of the Mississippi River in conflict (or denial) of French presence at Mobile, Biloxi, and New Orleans. "The hostile French, whose strategy of expansion and encirclement had been bemoaned by a legion of Carolinians had, with a few strokes, been removed form harm's way to the trans-Mississippi west."[58] As for the Spanish, one might well ask, "what Spanish?" as the name appears nowhere on the map, and St. Augustine is moved further south. What is printed on the map near today's Florida panhandle indicates "no Inhabitants from hence to the Point Florida."[59] With security assured, members of Parliament, investors and general mercantilists would also be happy to see the numerous rivers and coastal connections represented on this map.

Not long after the establishment of the colony and as the secretary of the Trustees, Martyn, sent a request to the surveyor, Samuel Auspurger, for "all the surveyed lands and accounts for whom the lands are, in what place, the number of acres, and the nature of the different soils."[60] For the next two decades these surveying efforts to describe cartographically the land for London eyes kept the colony in the mind of government officials who were continually reminded of the need to protect and support British colonial interests. A survey was announced in the 1753 *South Carolina Gazette* which would address the concerns Martyn expressed over a decade earlier and depict land possession and proprietors across South Carolina and Georgia.[61] The debate over the actions to take regarding French, Spanish, and even native encroachments put a burden on all the colonies, all of whom were asked to contribute supply in response. As the newest colony and perhaps still developing as a mercantilist unit in some conflict with South Carolina, there was some concern for London's attention. In the opening year of the Seven Years' War an article on Georgia appeared in the January edition of the *Gentleman's Magazine* in which the value of the colony was emphasized. "This colony . . . is thought of some consequence to the British interest, not so much for the value of what it may produce, as because it may serve as a southern frontier against any enterprize [sic] that may be formed by the French or Spaniards, if they should fortify themselves in Louisiana, as it is apparently their interest to do."[62] The following year Malachy Postlethwayt (d. 1767), a writer of economics, argued that this colony was of strategic importance to British success in North America. "[B]y the effectual support of this settelement on the one side, and Nova-Scotia on the other, we may effectually break the link of their intended chain . . . and therefore, this is matter of no less public attention than that of Nova-Scotia."[63] The trading axis (Saint Lawrence–Great Lakes–Mississippi) and the strategic axis (Quebec–Great Lakes–Mobile) were clear to both sides, but the territorial pretensions remained problematic and produced conflicting cartographical works that would become part of a debate in the interwar years between the War of Austrian Succession (1740–1748) and the Seven Years' War (1756–1763).

PAPER ENCROACHMENT TO THE NORTH: NOVA SCOTIA VERSUS ACADIA DISPUTE

In 1719 under the Regency government in France the young king, Louis XV, sent the Jesuit, Pierre-François-Xavier de Charlevoix (1682–1761), back to

New France to assess the boundary concerns regarding Acadia (in light of the Treaty of Utrecht, 1713) and to investigate an access route to the Pacific (i.e., the infamous Sea of the West).[64] Concerns had already been raised regarding English pretensions to lands beyond the French perceived boundaries of Acadia.[65] Although Charlevoix returned to France in 1722, his work, *Histoire et Description de la Nouvelle France*, was not ready to be published until 1744. When published, it was accompanied by maps produced at the Hydrographic Office (or Dépôt) by Bellin, *ingénieur-hydrographe* (hydrographical engineer), on the orders of the naval minister, Jean-Frédéric Phélypeaux, comte de Maurepas (1701–1781). Bellin (in effect, a geographer/hydrographer of the study) augmented Charlevoix's description with material found at the Dépôt such as navigational journals, astronomical observations, dispatches to the naval minister, and recent expeditionary reports from traders and explorers such as Pierre Gaultier de Varennes, sieur de La Vérendrye (1685–1749) and his sons.[66]

The third volume of Charlevoix's work began with Bellin's commentary on the maps included in the work as well as a brief state of the cartographical and hydrographical knowledge of the region. Through a review and analysis of the sources at his disposal, he suggested he had been able to produce a clearer geographical image of the region than had been provided to that point in time. While he wrote that he would not criticize earlier works that were on too small of scale to make them useful and others that were simply greatly flawed, he did present commentary on Popple's 1733 large map, *The British Empire in North America*. While more detailed than other contemporary maps and a comparative source for many of his geographical comments and locations, Bellin considered this well-regarded performance, on account of its size still as equally flawed as other extant works.[67]

As maps and charts often served strategic motives, Bellin also wrote, "One may be surprised not to find any soundings on my maps; I am aware that such details are extremely useful and I could have easily supplied them exactly, but I was prevented from doing so by particular reasons which have nothing to do with geography; as for the maps of ports which are not owned by France, I have indicated soundings."[68] The English translation of this work included an explanatory footnote of the "particular reasons," that is, "reasons of state" or so "foreigners might not know how to attack their ports."[69] This translation, an excerpt of Bellin's *Remarques*, appeared in the February 1746 edition of the periodical, the *Gentleman's Magazine* with Jefferys' modified copy of Bellin's map of the eastern part of New France included a month earlier in the January edition. Having been recently commissioned by Edward Cave (editor of the *Gentleman's Magazine*) to produce maps for the periodical, Jefferys' prominent career was just beginning. His work, *A new Chart of the coast of New England, Nova Scotia, New France or Canada . . .* , portrayed Nova Scotia on the peninsula where Bellin had

Acadie engraved.[70] The mainland region across which Bellin had *Nouvelle Ecosse* engraved appeared on Jefferys' chart filled, perhaps purposefully, with an inset map of the City of Quebec. This geographical controversy over what constituted Acadia and the extent of its boundaries had been brewing since the Treaty of Utrecht (1713) but was soon to blossom into a war of "paper encroachments."

Over the course of the decade that followed, France found itself in an increasingly tenuous position in North America and was forced to devote more of its energies to fend off attacks by the British fleets than to launch offensive naval assaults.[71] In 1745, the fort at Louisbourg on *Île Royale* (Cape Breton Island), which not only protected French interests in the Gulf of Saint Lawrence, but also was a major entrepôt for French trade between North America and the West Indies, fell to the British. Two subsequent attempts—including the 1746 naval disaster of the duc d'Enville's armada—to restore the fort to French control were unsuccessful.[72] The Treaty of Aix-la-Chapelle (1748) provided land restoration (e.g., the fort at Louisbourg was returned to French control) for the major powers, but solved little and only temporarily established a truce.

In 1750, Roland-Michel Barrin, marquis de La Galissonière (1693–1756), the former interim governor of Canada (1747–1749), became the director (*garde du Dépôt*, 1750–1756) of the Hydrographic Office and was named to the Anglo-French Boundary Commission, whose task was to settle boundary disputes in North America.[73] His naval and diplomatic experience taught him not only the significance of a strong French presence in North America to frustrate further English ambitions, but also the importance of accurate maps and charts of the North American coast, specifically the Canadian coast and waterways. He recognized that with the uncertain peace following the War of Austrian Succession and the rise of a strong English naval fleet, French overseas possessions were vulnerable as was the position of France on the continent.

In his report, *Mémoire sur les colonies de la France dans l'Amerique Septentrionale*, from December 1750, La Galissonière adopted an urgent tone on English encroachments and the security of Canada. He acknowledged that the protection of the region was a burden for France, but it was of such strategic value to thwart English ambitions in America and to prevent English superiority in Europe.[74] To challenge those ambitions France needed greater geographical and hydrographical knowledge of the region. In preparation for the inevitable maritime encounter in North America, he organized the expedition (1750–1751) of the naval commander and hydrographer, Joseph-Bernard de Cogolin (1724–1805), marquis de Chabert, whose 1746 expedition had already provided observations along the coast of Acadia that improved French charts. In 1750 "as an agent of the Colonial Machine,"[75] Chabert once again set off to the islands, coastlines, and waterways of eastern Canada, and knowing that he would encounter the British was instructed by the crown to show his orders "presuming that they would not be opposed to work which had the objective of the common safety of all navigators."[76] Upon his return in late 1751, Chabert began under La Galissonière's direction to prepare his hydrographical survey for a publication (in 1753) to include his journal, charts, and measurements and astronomical observations. His work provided valuable information for all navigators who had the occasion (or necessity) to voyage in the dangerous waters of this region. In the supplement to the 1754 edition of the *Gentleman's Magazine*, Chabert's work was listed among the "best foreign books published."[77] In the 1755 March edition of this periodical, the latitude and longitude values from Chabert's work were reported with the understandable correction of longitude to St. Paul's in London. More interestingly, however, given the on-going dispute were the location descriptions that had Halifax in Nova Scotia (not Acadia as reported in Chabert's work) and Cape Sable Island at the Western extremity of Acadie or Nova Scotia in direct opposition to the French position that Acadia was a distinct region, not interchangeable with Nova Scotia and only the peninsula region at that.[78] A situation that was not surprising given the reports that came out of the Anglo-French Boundary Commission.

In London in 1749, the secretary of state, John Russell (1710–1771), duke of Bedford, appointed William Shirley (1694–1771), then governor of the English province of Massachusetts Bay (1741–1749 and 1753–1756), as commissary to the Boundary Commission because of his staunch position against French encroachments. Shirley gathered information from his American correspondents and various government offices in London but was not sure that a commission was the proper way to settle such an imperial problem.[79] With the meticulous and unremitting attitude of a lawyer, Shirley intended to uphold the position of Bedford's expansionist faction which was problematic for his patron, the duke of Newcastle, who championed the faction for peace and settlement.[80] Over the course of the interwar years these internal factions vied for dominance while the two states moved to solidify boundaries and settlement.[81]

The initial meeting of the commissaries at the end of August in the Paris apartments of La Galissonière served only to bring the gentlemen face-to-face to politely clarify for one another the intransigence on both sides. The opening statements which followed in September and November by the commissaries put in text how far from agreement they were. Such an occasion provided *good copy* for the *Gentleman's Magazine* which contributed to the rhetoric of indignation over French pretensions in the Americas. For example, months before the commissaries met, in the February edition, the editors provided a commentary on the French and the task ahead for the English commissaries. The French will "endeavour to stretch their pretensions, and allow us little more than the peninsular. . . . It should seem indeed, that they have been meditating encroachments for some years past, for their geographers have greatly enlarged their own, and in proportion lessen'd our boundaries." Worse yet, "our map-makers, and some of them in royal pay too, have as stupidly followed without a meaning, as the French have meaningly contrived them. I make no doubt, that the descriptions of our own copying geographers will be produced and insisted upon against us." With the interest of this new colony in mind, this author hoped the commissaries would utilize the "experience of the board of trade" in their negotiations in the hopes "to catch the French in their own snare," but "avoid making any concession towards the sea that may facilitate the navigation of the French."[82]

Months later as the commissaries in Paris were making little headway, in an account of Nova Scotia in the June 1750 edition of the *Gentleman's Magazine*, the author expressed the English concerns regarding French encroachment. South of the entrance to the St. Lawrence River was Gaspé Bay, "a deep and good harbor."

> Here the French, contrary to treaty, have continued to carry their fishery, and pretend to assume to themselves a right over the country behind it, which they distinguished in their modern maps by the name of *Gaspesee*. This name they do indeed sometimes extend so far, as to take off the greatest part of Nova Scotia, and leave us little, if any thing, more under the title of Acadia, than the peninsula. Such a *paper incroachment, if not well attended to, may in time be construed into a sort of claim by prescription.*[83]

Commissaries on both sides were well aware of the economically and politically strategic location of Nova Scotia, but English commissaries traveled to Paris with the voices of critics in their ears. Not only had their geographers "stupidly" followed the French thus allowing a "paper incroachment," but Britain had done little to counter the French influence in the region. With the loss of Port Royal in 1710, Acadia fell to English control, but as the author of the piece on the "Importance of Nova Scotia" reported, there was little British settlement to challenge the influence of the French in the area.

> I have been more particular, to shew why the French may at a congress claim this province, of which they have *actual possession*, as an equivalent for Cape Breton. And if one must be given up, it is beyond all dispute that Nova Scotia is worth more to France than that island and Newfoundland together, considering its extent, cultivation, timber for ships, masts, mines, furs, provisions, multitude of fine harbours, and fishing-banks, which give place to none in the world.[84]

Such a bountiful land could not be allowed to fall under French influence of any sort, economic, political, and even religious. In the absence of "*actual possession*," Britain had to document the boundaries in text and counter contradicting graphical imagery with other images or words that would place France on the defensive.

The mémoires (or memorials) produced from this Anglo-French boundary dispute illuminate the curious, mutually acknowledged distrust of maps,[85] alongside the use and discussion of them despite claims of their inherent unreliability. In their memorials dated 11 January 1751 and 23 January 1753, the English claimed only to cite maps to correct French assertions and because the French gave greater credit to maps. "Maps are appeal'd to by us only in answer to the Assertion, that Charts of all Nations confine the Limits of *Acadia* or *Nova Scotia* precisely to the *Peninsula*."[86] Furthermore, in their reply to the French memorial of 4 October 1751, the English repeated, "In Treating of Maps it may not be improper to remind the French Commissaries, that the Commissaries of the King of Great Britain were not the first who appealed to these as an Authority in the present Discussion: That they never cited them but to correct Mistakes made by the French Commissaries."[87] In the English memorial of 1753, they argued again that their appeal to maps was only in response to the French as they [i.e., the English] had already, "disclaimed any very great Reliance upon the Evidence of Maps, even where they have proved them to favour the Claim of the King of Great Britain. And that they should not at this Time have gone into a more minute Consideration of them, if the French Commissaries had not made it necessary, by again giving a much greater Credit to Maps than they deserve, and by affecting to make them seem material in the Discussion of the Point before us."[88]

Regarding the uncertainty of maps, the English commissioners wrote, "maps are from the Nature of them a very slight Evidence, Geographers often lay them down upon incorrect Surveys, copying the Mistakes of one another; and if the Surveys be correct, the Maps taken from them, tho' they may shew true Position of a Country . . . can never determine the Limits of a Territory." In fact, the English suggested that the kind of weighty evidence that should be used to produce maps "would be themselves a better Evidence," and should be the material put forward in such a dispute. "In this particular case, histories as well as maps, are most uncertain guides."[89] The French commissioners agreed as well that the authority of geographers ought not to be decisive, as they were more concerned to give their works the air of system and truth as well as an appearance of science and research than to fix the rights of princes and the true boundaries of regions.[90] And yet, still, maps were part of the discussion. The French even challenged the English to bring forward "ancient" maps to support their position.

Feeling compelled to leave no comment by the French unaddressed, the English commissaries sallied forth for several more pages on the evidence brought to bear by "more ancient" (i.e., seventeenth century) maps and at one point concluded, "that the most authentick antient Maps made in different

Countries are in Favour of the Demand of Great Britain: and that the French Commissaries were a little misled when they called upon us to produce such antient English maps."[91] "Among the great variety of maps, which we have produced of different countries made at so many different periods of time, there is not a single one to be found, not even among the French maps themselves, which does not expressly confute the main and the essential part of the system of the French commissaries."[92]

For four years the Boundary commissioners had traded memorials supporting their territorial claims. In brief, though they were not, the French claimed by ancient boundaries and the wording of the Treaty of Utrecht that Acadia was only a strip of land along the southeastern coasts of the peninsula, but they were willing to concede that the entire peninsula was ceded to the English. Alternatively, the English claimed a much larger region called either Nova Scotia or Acadia as the land extending from the Saint Lawrence River along the western boundary, to the Kennebec River along the southern boundary (as the land between the Penobscot and Kennebec Rivers was claimed in the past by the French as part of Acadia, but already in English hands), to the Chignecto Isthmus at the northern boundary and to the peninsula at the eastern boundary. For the English, any suggestion that the Treaty referred only to the peninsula was sheer absurdity. From the French perspective, any interpretation that entertained claims on the mainland extending from the Saint Lawrence River clearly encroached on the well-established lands, that is, French lands, of *Nouvelle France* or Canada. Their lengthy debate and verbose assertions had serious economic and military implications, as claims of the mainland would provide the French not only with "access to the fur-bearing animals of the interior but also to a river route via the St. John River from Quebec to the Atlantic, a vital link during the winter months when the ice prevented navigation in the St. Lawrence."[93] For the English, while control of the peninsula would provide command over the cod fishing industry off the coast, they "coveted the St. Lawrence not only as a communication route but also as a natural boundary between New France and New England."[94] The English position was clear for at least a decade: Nova Scotia was valuable real estate to both countries and its full geographical extent simply could not be allowed to stay in French possession.

THE MAPS OF THE MEMORIALS

Its geographical extent and the alleged "paper encroachments" demanded the presentation of maps by both sides. Two maps discussed in the memorials of the commissioners were Bellin's 1744 map of the eastern part of New France dedicated to Maurepas, *Carte de la Partie Orientale de la Nouvelle France ou de Canada*, and Henry Popple's 1733 map, *Map of the British Empire in*

North America.[95] On Bellin's map, Acadia was depicted as the peninsula, while the land extending eastward from the Saint Lawrence River to the Bay was Nova Scotia. Even his *Carte de l'Amerique Septentrionale* also from Charlevoix's *Histoire* (1744) displayed the same territorial labels. From Bellin's works, then, the French perspective appears to be that Nova Scotia and Acadia were two distinct regions with Acadia as the peninsula. Popple's 1733 map provided a similar presentation with Acadia as the peninsula and Nova Scotia as the land from the Saint Lawrence River to the Bay, but Popple's labels were not parallel as on the Bellin maps. The territorial labels were perpendicular—such that the positioning and spread of the letters of Nova Scotia extending to the peninsula on which Acadie was engraved left intact the English interpretation that these were not distinct regions.

Ironically, but in accordance with their nationalistic interpretations, each of these maps would be used by their opposing boundary commissioners to challenge the disputed claims. Bellin's map was touted by the English as evidence that Acadia was part of Nova Scotia and an officially sanctioned map. "Mons. *Bellin's* Chart, was with other Plans, composed by special Directions from the Marine of *France*."[96] To which the French replied not so much in favor of Bellin's 1744 map because of its depiction of a distinct and independent region of Nova Scotia, but because he had placed Acadia on the peninsula alone in agreement with French claims. After all, the important element in the argument, as emphasized by the French, was not the labeling of Nova Scotia (or the mainland area), which the French preferred to view as just part of *Nouvelle France* and therefore not part of the discussion, but the labeling of the peninsula.[97] Perhaps, Bellin himself might have argued that given the title of the map, *Carte de la Partie Orientale de la Nouvelle France ou de Canada*, it was overreaching for the English to suggest that the absence of a label for "New France or Canada" marked "on the Southern Side of the River St. Lawrence" provided evidence that the French did not view that area as part of *Nouvelle France*.

For cartographic evidence, the French turned to the map of Henry Popple not only as an example of an officially sanctioned map by the English government, but also in support of Acadia as only the territory of the peninsula. Alured Popple, Henry's brother, was the secretary of the Board of Trade and may have assisted his brother in his efforts to produce a map of North America.[98] Remember, Popple had emphasized in an engraved text block on his map the nature of the map's origins and accuracy declaring that he had the approbation of the Lords Commissioners of the Board and that he had access to the "authentic records and actuals surveys transmitted to their Lordships."[99] The English denied French claims that Popple's map was officially sanctioned by the Commissioners, "who might very well approve *of such an Undertaking* but who never superintended or approved of Mr. *Popple's* Manner of executing it. Mr. *Popple* inserted this marginal Note *meerly to secure a better Reception to his Work* [my emphasis]; . . . the Board of Trade at the Time gave it no extraordinary Sanction."[100] To accord his work "a better reception" then, Popple suggested that it had a more substantial approbation. Both sides could argue that "paper encroachments" were made for the very same reason, "a better reception."

> English commissaries also questioned the usefulness of the map for French claims: Mr. *Popple* has marked the Peninsula with the Name of *Acadia*, and the whole Country westward as far as the Southern Bank of the River St. *Lawrence* with the Name of *Nova Scotia*, of which he makes St. *Croix* the western Boundary, which shews he thought the Country of *Acadia* or *Nova Scotia* extended from the southern Bank of the River St. *Lawrence* to St. *Croix* and makes his Map but of very slight Authority for the *French* Commissaries, who confine *Acadia* or *Nova Scotia* to the south-eastern Part of the Peninsula.[101]

For the French, the depiction of Nova Scotia on the mainland fit well with their understanding that it was part of *Nouvelle France* and, again, as they had emphasized for the Bellin map, the peninsula was restricted to Acadia, the central issue with which they would conclude their *mémoires*.[102]

The English, on the other hand, in the conclusion to their discussion of maps in their final memorial wrote that "if the French Commissaries were admitted to decide the antient Limits upon the Testimony of Maps, they have not yet found any one Map made in any Country in any Time which assigns the antient Limits according to their Pretensions."[103] Given the dizzying retorts in the published *mémoires* on both sides, one can very well understand why the face-to-face meetings ended, but equally the twists and turns of the arguments and interpretations of pieces of evidence (the least of which were maps) brought forward indicated so clearly that an agreement at the table or in the press would be most unlikely. It was only a matter of time as to when war would erupt.

POST-MEMORIAL MAPS

In 1753 when successful negotiations were looking bleak, Bellin's contemporary and well-known cartographer, Didier Robert de Vaugondy (1723–1786), dedicated and presented to Marc-Pierre de Voyer de Paulmy, comte d'Argenson (1696–1764)—Louis XV's minister of war and secretary of state—a map of Canada which by its dedication and depiction of the English and French possessions, "nearly provoked an international incident."[104] Since the English viewed Acadia (or Nova Scotia) as the land extending from the Saint Lawrence River up to and including the peninsula, upon which (and only here) Vaugondy had inserted the name Acadia as a strip of land along the southern coast of the peninsula, the English commissioners "sniffed French propaganda" and raised voluble concerns that their treaty possessions had been shrunk not just to a mere peninsula, but to a strip of land on the peninsula.[105] Worse yet, this map appeared to have official sanctioning. Not only had the dedication to d'Argenson given the map the air of officiality, but also Vaugondy's bold advertisement for the map in the *Mercure de France* suggested that he was presenting a map with detail not seen before and, in particular, for Acadia had based his presentation on reports communicated to him from the Ministry.[106] Vaugondy's promotional efforts had backfired, and, in the following month of September, he issued a retraction clarifying and emphasizing that he had not been privy to any official reports from the Ministry, but he did leave it to the public whom he acknowledged had his work and those previous works in hand to judge the quality of his work.[107] In a similar situation to Delisle's earlier work on North America (1714), this map participated in the rhetoric of power and the discourse of science as Vaugondy suggests by his advertisement that the quality of information from an officially sanctioned map carried more certainty. Since the map *produces* the territory, Britain had to respond.

In 1754, Thomas Jefferys printed a work in support of the English stance on Nova Scotia. *The Conduct of the French with regard to Nova Scotia; From its first settlement to the present time. In which are exposed the falsehood and absurdity of their arguments made use of to elude the force of the treaty of Utrecht, and support their unjust proceedings.* The author of the piece, Braddock Mead (1688–1757), also known as John Green, took up the verbal cudgel left behind by William Shirley. It was sheer absurdity for the French to think to "wrest" from the English people a "spacious province" that they [the French] had, in effect, "invaded" pretending "that we are entitled to no more than the small share they have left to us, which is at most the peninsula or southern division of *Nova Scotia*." As to the role of the mapmakers in this charade, Green writes:

> Their geographers and historians have been influenced to prostitute their pens in the most shameful manner, to serve the injurious cause; and their principals, who set them to work, have not been ashamed afterwards to make use of such corrupt evidence, consisting of the lowest chicanery and most barefaced falsifications, as the chief, and in effect the only arguments on which they ground their pretensions. In short, their reasons are so consummately fallacious, inconsistent and trifling, that their defence of the injury ought to be taken for sneer, and is no less provoking than the injury it self.[108]

He calls into question not only the methodology of these geographers but the influence of power on map making. The French are using knowingly "corrupt evidence," and prostituting the practice of geography to the interests of power rather than the interests of science.

Green put before a popular audience what had already passed before the eyes of English and French officials. On Bellin's 1744 map of the Eastern portion of New France, Nova Scotia was labeled on the mainland. He pointed out that Bellin was no "ordinary geographer," but the "hydrographer to the marine, as well as censor royal" so if he contradicted a source given his position, his work would hold more weight. Furthermore subsequent maps by Bellin which omitted the name of Nova Scotia were not so much a correction of a mistake in the earlier 1744 map, but a political decision. "[I]t was not, as may be presumed, in consequence of being better informed, but because he was otherwise directed or inclined."[109] Green suggested that Bellin was instructed by the naval minister to label his maps in accordance with French pretensions. "Our neighbours are very dexterous at either expanding, or contracting; for, whenever they please, they can turn a single fort into a large country, and reduce a large country into a piece of coast."[110] These were calculated decisions to publish maps that defined a territory. In other words the boundary or identification of a particular area was purposefully adjusted to fit the political interests of the state. French map makers were culpable in the dubious politically-charged geographical machinations of the French state.

Green further implicated French geographers who were in service to the state.

> 'Tis true, that Mr. D'Anville, in answer to a charge of marking the bounds of some British dominions in America amiss, expresses a surprise 'That any body should imagine a thing of this kind done by a geographer, could be either of prejudice or advantage to the rights of crowned heads.' I [Green] surprised at it, no less than he; for it would be strange indeed, if the bounds of kingdoms, any more than the situations of places, were to depend on the arbitrary will of the geographers: that would be to have kingdoms at their disposal. But then, I see it has been the case; and at this instant the maps but just now mentioned are produced as arguments, to support the French allegations. . . . there is no extravagant demand or assertion; no inconsistency or chicanery, within the compass of invention, which the French may not be capable of having recourse to, when they have any favourite point in view.[111]

Green's assertions so well illustrate the well-understood paradox of these socially constructed texts that on the one hand were championed as clear works of science—hence trusted representational works—and on the other as convenient inventions, that is, purposefully part of a discourse of power. Territory is negotiated and constructed, and maps were the vehicle of this construction.

In the year following 1755, Green provided Jefferys with a map of Nova Scotia and an explanation to accompany its publication, *Explanation for the New Map of Nova Scotia and Cape Britain with the adjacents parts of New*

England and Canada.[112] A map viewer was not to regard his map as an "invention"—but a *true* representation—as he listed on the map those observations used to geographically situate the information, for example, latitude and longitude values for Cape Sable taken by Chabert in 1751. He also noted on the map "the French Incroachments and extensive claims" put forward on the maps by d'Anville, Robert de Vaugondy, and Bellin. In the *Explanation*, Green's rhetoric against French pretensions was more toned down, and he even extended some praise for France's attention to improving charts of the region. He was, however, quite pointed in his remarks toward navigators and geographers in general. For example, he reproached navigators and geographers for practices that did not advance science. Navigators were well aware of the "wretched condition of charts," but seemed "to know no utility in observing latitude" and neglected "to make exact observations at all ports and headlands."[113] Geographers concealed their sources from one another, and if they would but "communicate them, the maps resulting from their several labours would be more accurate and complete, if not perfect; nor would they differ so enormously, if they did not altogether agree."[114] French navigators such as Chabert, who had crown support for his "long and hazardous voyages" had also urged pilots to make observations for the purposes of improving charts of the region. Green wrote: "He speaks with respect to the sea-officers of his own nation; and it could be wished, that what he says of them might with equal justice be applied to ours." Green despaired that the recognition in France of the correlation between navigation and commerce "seems not as yet to be sufficiently perceived here, . . . that Science is not the object of ministerial care, as in France."[115] The Colonial Machine that France had developed beginning in the late seventeenth century was still about two decades away in Britain. Green might have argued nonetheless that maps could be *truer* representations if only navigators and geographers performed their responsibilities more conscientiously. After all France had the infrastructure, but Green would suggest that its geographers erred by serving the interests of power rather than the interests of science.

Appearing in the same year, Bellin's map of North America (1755) represented "Acadie" on the peninsula while the mainland area was not identified as Nova Scotia, but simply part of Nouvelle France (New France)[116] (see figure 4.1). In his accompanying *Remarques*, Bellin affirmed the French position that "Acadie" was the southern portion of the peninsula and that this region was ceded to the English by the Treaty of Utrecht (1713). Nouvelle Ecosse (Nova Scotia) appears neither on the map or in the *Remarques*. As for the debate, he directed readers to the *Mémoires of the French Commissaries.*[117] Perhaps in response to Green's comments or the recent publication of John Mitchell's map of North America (see below), Bellin summed up his philosophy citing a line from Persius's *Satire* IV (line 42): "Caedimus, inque vicem prebemus crura sagittis," that is, "We shoot, and in turn present a target for the barbs of others." This line appeared in the first few pages of his *Remarques* in which Bellin reflected on the lack of accuracy and detail for maps of North America from the seventeenth and early eighteenth century.[118] While he wrote that one might be surprised by the variation in the extant maps, their authors should be excused because of the lack of knowledge of the area, a situation that was changing only in the past decade. Bellin was hopeful that the field was improving and sites among the best works those from d'Anville, Robert de Vaugondy, Green, and John Mitchell's map (see below). Adopting a propitiatory tone, he suggested that it was better to give these geographical works their due and to use them as a "flame" to light the way to more commendable works. Simply targeting works that used the available information only exposed one's own work to "the barb of others." Bellin continued his instruction suggesting that one must avoid such a stumbling block and that being for the nation or for a seductive idea could make one sacrifice knowledge. Geographers must rise above nationalistic debates and provide a sound methodology for their productions.[119] Despite the *talk* on both sides, this debate was not about accurate scientific information. Both sides knew that maps defined territory and used or revised them and praised or denounced them. For the issue of the peninsula, an accurate survey by triangulation would not have solved anything as the question was not *where* is the boundary of Acadia *on* the peninsula but that the entire peninsula was Acadia (and therefore Nova Scotia). While these maps participated in the discourse of science, they more readily involved the rhetoric of power, but in the eighteenth century the power discourse embraced the rhetoric of scientificity, and maps served the state as authoritative instruments of statecraft, especially in colonial disputes.

Connections to institutions responsible for knowledge of the colonies could provide a good foundation for cartographic works of America. In 1749 George Montagu Dunk, Earl of Halifax, the new president (1748–1761) of the Board of Trade, made the acquaintance of John Mitchell, the Virginia physician and botanist who had retired to Britain for health reasons, and soon asked him to compile a map of North America.[120] What Halifax and Mitchell discovered was the inadequacy of the Board as a source of cartographic information. At the same meeting in which the Board reaffirmed the instructions for the commissaries, William Shirley and William Mildmay, Halifax and the Board ordered that a circular letter be sent to all colonial governors for "as exact an Account as you can of your Limits and Boundaries of the Territory [of your province or colony], together with a Chart or Map."[121] Completed in late 1754 and supporting Halifax's political stance to deny the French any territory south of Canada or east of the Mississippi River, Mitchell's eight-sheet map of dominions, *A Map of the British and French Dominions in North America*, proposed to correct the many errors from pretended authorities[122] (see figure 4.2).

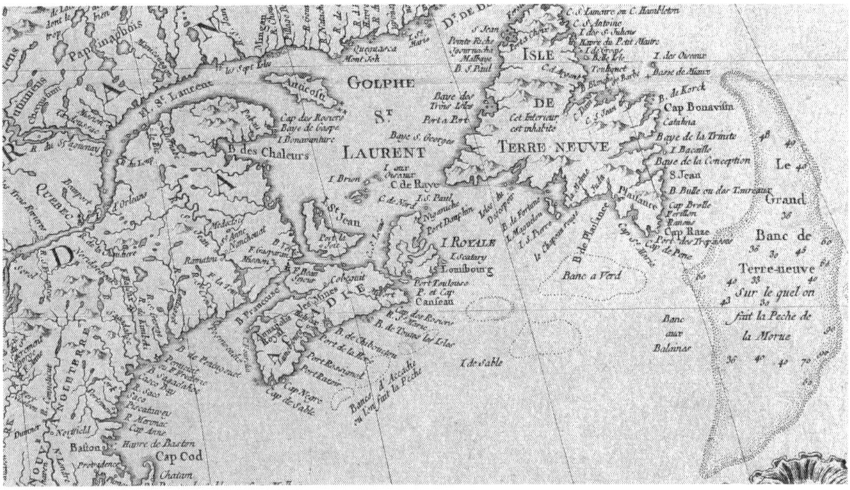

Figure 4.1. Detail from Bellin's *Carte de l'Amerique Septentrionale* (1755). *Source:* Courtesy of the Beinecke Rare Books and Manuscript Library, Yale University

Mitchell's map exuded scientific and political authority. He included a list of latitude and longitude measurements "observed and deduced" with brief explanations. Nearby on the map was a lengthy engraved text indicating that this map originally laid down in 1750 was revised and improved by recent sources including the most recent observations by Chabert. As for the region of New England and Nova Scotia, Mitchell reported that recent maps and charts were greatly flawed having relied on a "feigned survey."

> The only authority they have for all this is a feigned survey by a pretended surveyor general Blackmore in 1711–12 . . . Mr. Moll published it as an actual survey 'made by her majesties especial command' from which this coast has been thus erroneously laid down ever since. . . . In short we do not find a single spot hardly justly laid down in these our New Maps of Nova Scotia. . . . . We are so far from improving them in the Geography of America that we see it made worse & worse for want of certain observations, which we have endeavoured to collect & thus to represent in one view.[123]

Mitchell intended that, unlike previous maps, this work would provide a more authoritative view based on certain knowledge.

Most of one of the sheets was taken up with the elaborate cartouche carrying the royal arms and images of the abundance of the region, that is, maize, palm trees, fish nets, and a beaver. A native kneels in supplication looking upward towards the British coat of arms as the seated figure of

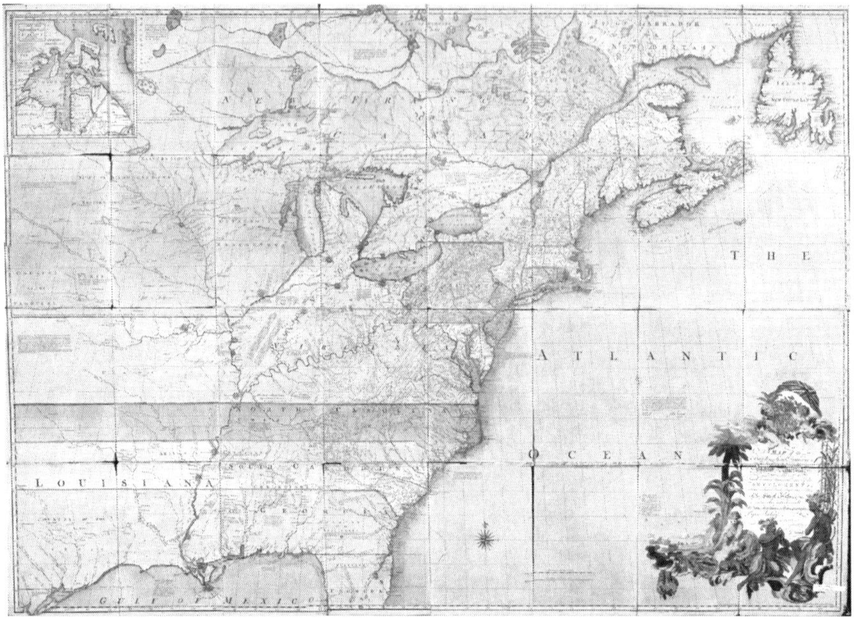

Figure 4.2. Mitchell's *Map of the British and French Dominions in North America* (1755). *Source:* Courtesy of the Yale University Map Library

America (a native woman) also raises her eyes upward with her right hand on her breast in a show of deference as the native element looked to Britain for protection. In the distance along the left hand side of the cartouche was the scene of a peaceful harbor. The symbols used depicted towns, forts, settlements, roads with distances in miles, and deserted Indian villages, etc. Also on this sheet, perhaps in response to the accusations made against Popple's map, was the statement: "This map was undertaken with the approbation and at the request of the Lords Commissioners for Trade and Plantations; and is chiefly composed from draughts, charts and actual surveys of different parts of his majesties colonies & plantations in America; great part of which have been lately taken by their lordships orders, and transmitted to this Office by the Governors of the said colonies and others. John Pownall, secretary. Plantation Office Feb. 13, 1755." In contrast to Bellin's map of the same year, the French were denied any lands east of the Mississippi River, and in support of the English point of view, Nova Scotia or Acadia were written across the disputed region in indication that the names were interchangeable.

Halifax "unveiled" Mitchell's map in response to what he perceived as "severe territorial concessions" being considered by Sir Thomas Robinson negotiating a compromise with the French "for the creation of a neutral zone in the Ohio valley."[124] About forty years after Bobé had written to Delisle (1714) about Raudot's concern over the representation of boundary lines that pushed French territorial possessions into the lands of other countries, the Lord Chancellor, Philip Yorke (1690–1764), Earl of Hardwicke, expressed unease with the limits of British colonial possessions. "I fear very inconvenient Consequences from it [the map], for it carries the Limits of the British Colonies as far, or farther than any other, which I have seen. If it should come out just at this juncture, with the supposed Reputation of this Author, & the Sanction of the Board of Trade, it may fill people's heads with so strong an opinion of our strict Rights, as may tend to obstruct an Accommodation." The map's printing was not delayed. Edney suggests that "Halifax's intention in having Mitchell's map published was to create a definitive statement that would shape the conceptions of politicians and public alike."[125] In this role, the map was successful as "[t]he British ministry abandoned a compromise with the French and soon had plans in hand to oust the French from their colonial encroachments by force."[126]

In the first year of the Seven Years' War and shortly after the publication of the *Memorials* from the Boundary Commission, Jefferys published *Remarks on the French Memorials Concerning the Limits of Acadia* in which he enclosed two maps one according to the French position and one according to the British.[127] The French map clearly limited the region of Acadia (à la Bellin) to a portion of the peninsula, while the English map depicted Acadia and Nova Scotia as interchangeable geographical labels (à la Mitchell). Jefferys cautioned that maps were able to convey an idea more quickly than text. "It remains only to take notice of the two maps prefixed to this treatise; about which, however, little more need be said than what appears on the face of them, where they carry every thing which may be necessary for their explanation, and to shew their use." In contrast to the claims by the commissioners, Jefferys intended to make the English case using maps. In fact, "[a]s the French thought fit to insert one in their Memorials to illucidate their system, I judged it proper to contrast it with another which exhibits the state of matters, according to the authorities produced by the English. Maps, to be sure, give a more distinct view, as well as convey a more quick idea, of things than writing; by combining the several matters scattered in discourse, and compendiously exhibiting as in a picture the whole at one view." As "sensible images" maps were powerful tools of government. "I am glad, therefore, that the French have set an example, which gives us an opportunity of reducing the controversy to a narrower compass, and at the same time impressing the force of our evidence by sensible images."[128] The challenge remained as both sides claimed to present "sensible images" upon which there was little agreement. The inability of the Anglo-French Boundary Commission to resolve the territorial dispute of Nova Scotia, thc increase in rhetoric of encroachment on both sides, as well as the events in North America (and on the continent) made it all the more evident that the power struggle would be determined by naval dominance, successful territorial occupation, fort constructions, native alliances, and, ultimately war.

CONCLUSION

These "paper wars" between Britain and France were brought to an end by the Treaty of Paris (1763), by which France lost her colonial possessions on mainland North America. As a consequence with respect to mapping, the tables were now turned as France had to rely on British maps (and charts) to gather information on the expanding colonial infrastructure as well as coastal and some interior geographical data.[129] In the period after 1763 and before the War of American Independence, Britain moved to harness the talents of their surveyors such as Samuel Holland and J. F. W. Des Barres. However skilled they were, Stephen Hornsby suggests that as Britain "cobbled together survey teams from existing resources" the shortcomings of the lack of organization became clear and contributed to the decision to establish the Admiralty's hydrographic office which still did not take place until 1795.[130] While the efforts of these surveyors better served the British navy than their land forces in the war against the colonists, it helped to lay the foundation for a more systematic approach and to highlight the need for a directing institution. These efforts at a systematic survey based on triangulation along America's Atlantic coastline set the stage for Britain's Imperial Machine, as "[t]he geography of empire was placed on a rational, scientific basis."[131]

Decades before these scientific surveys created overlapping large scale cartographic documents (which did not necessarily quell debate), small scale maps by French and British map makers participated in the continuum of political posturing seen in the decorative elements discussed in chapter 1. The maps themselves (sometimes with supporting imagery) defined the territorial dominion of the British or French state. The recognition that maps produced territory and their appeal to the discourse of science made these powerful instruments of the state. They would be made public (sometimes with accompanying printed remarks or advertisements) as part of a colonial agenda and subsequently championed, retracted, or disavowed depending on the political environment. Empirical data was still rather slight for these colonial regions, a *wild west* of sorts for cartographical representations, but the state and consumers were thirsty for such productions and map makers like Bellin and Green/Jefferys were eager to tout their expertise often at the expense of the other. There was in the remarks of Bellin, Green, and Chabert a sense that conscientious efforts by geographers and navigators could produce maps and charts with certain knowledge. Certainly maps participated in the rhetoric of power and a posture of dominion in the colonies but like the regional maps discussed in chapter 2 whether they presented the lay of the land, the area of a disputed boundary, or the topographical challenges for public work projects, these maps were useful tools for the administration of the state—the British state, the French state, and the emerging independent entities of colonial America. As representational documents trusted maps, that is, based on scientific methodology, would continue to denote ever-increasing certain geographic data. In the next chapter, the story takes us to the East Indies and efforts by Britain and France to chart dangerous and unfamiliar seas in the hopes to reach the level of certain hydrographic data.

NOTES

1. Cecil Headlam, ed., *Calendar of State Papers, Colonial Series, America and West Indies, 1699* (London: H. M. Stationary Office, 1908), 66, 17 February, #108, "Council of Trade and Plantations to Mr. Secretary Vernon, 17 February 1699."
2. Headlam, *State Papers, 1699*, 66.
3. Headlam, *State Papers, 1699*, 604-5 [Addenda 1621–1698], 28 October 1687, #1156, "French Ambassador to the King."
4. Headlam, *State Papers, 1699*, 604 [Addenda 1621–1698], 20 May 1687, #1153, "Extract of a letter from Governor Thomas Dongan."
5. James E. McClellan III and François Regourd, "The Colonial Machine: French Science and Colonization in the Ancien Regime," *Osiris* 15 (2000): 32.
6. James E. McClellan III and François Regourd, *The Colonial Machine: French Science and Overseas Expansion in the Old Regime* (Turnhout: Brepols, 2011), 485.
7. John Green, *Remarks in Support of the New Chart of North and South America* (London, 1753), 3.
8. John Pickles, *A History of Spaces: Cartographic Reason, Mapping and the Geo-coded World* (London & New York: Routledge, 2004), 5
9. Matthew Edney, "Reconsidering Enlightenment Geography and Map Making: Reconnaissance, Mapping, Archive," in *Geography and Enlightenment* ed. David N. Livingstone and Charles W. J. Withers (Chicago: University of Chicago Press, 1999), 169.
10. Some of the material for this section is drawn from my first book, *When France was King of Cartography: The Patronage and Production of Maps in Early Modern France*, 101–6. The Delisles' attention to mapping the Americas is evident in the three *portefeuilles* from the Delisle archive covering America, AN, Marine, 2JJ 55–57, *Delisle Archive*.
11. Guillaume Delisle, *L'Amérique Septentrionale dressée sur les observations de Mrs. de l'Académie Royale des Sciences & quelques autres, & sur les memoires les plus recens* (Paris, 1700), BN, Cartes, GE BB 565 (XIV) 67.
12. Petto, *When France was King*, 124, n. 4.
13. Jean Delanglez, "Franquelin, Mapmaker," *Mid-America* 25 (1943): 36–37.
14. Marcel Giraud, *A History of French Louisiana*, Volume One: *The Reign of Louis XIV, 1698–1715*, trans. Joseph C. Lambert, revised and corrected by Giraud (Baton Rouge: Louisiana State University Press, 1974), 7.
15. Jean Delanglez, "The Sources of the Delisle Map of America, 1703," *Mid-America* 25 (1943): 276–80; Petto, *When France was King*, 102. Claude had communicated with members of the expedition team, for example, M. de Beaujeu, M. d'Amanville, and Jean Cavelier, René's brother. De Beaujeu would later provide a manuscript map of the Gulf of Mexico. The Delisles benefited not only from another source, Henri Joutel's journal written during the La Salle expedition but also from their later correspondence on the location of various rivers.
16. Nelson-Martin Dawson, *L'Atelier Delisle: L'Amérique du Nord sur la table à dessin* (Sillery, Québec: Editions du Septentrion, 2000), 167.
17. Louis Hennepin, *Nouvelle Découverte* (Utrecht, 1697), preface.

18. See BL, Lansdowne Ms. 849. Various letters and state papers from the late seventeenth century regarding French encroachment.

19. Cecil Headlam, ed., *Calendar of State Papers, Colonial Series, America and West Indies, 1708–1709* (London: H. M. Stationary Office, 1922), 421–24, 10 July 1709, #632. Also see Louis De Vorsey Jr., "Maps in Colonial Promotion: James Edward Oglethorpe's Use of Maps in 'Selling' the Georgia Scheme," *Image Mundi* 38 (1986): 35–45.

20. Cecil Headlam, ed., *Calendar of State Papers, Colonial Series, America and West Indies, August 1714–December 1715* (London: H. M. Stationary Office, 1928), 116, 11 March 1715, #271.

21. Dawson, *L'Atelier Delisle*, 165-171; Delanglez, "Sources," 280. The two maps presented were Guillaume Delisle, *Carte du Canada ou de la Nouvelle France* . . . (Paris, 1703), BN, Cartes, GE BB 565 (XIV) 89 and *Carte du Mexique et de la Floride* . . . (Paris, 1703), BN, Cartes, GE BB 565 (XIV) 93.

22. For a description of Delisle's correspondence with Bobé as well as Bobé's information network, see Dawson, *L'Atelier Delisle*, 109–16. The correspondence between Delisle and Bobé can be found in AN, Marine, 2JJ 56, X, 26; AN, Marine, 2JJ 60 XVI; AN, Marine, 2JJ 91, Section 7. See also Anonymous, "Curious Correspondence of De L'Isle the Geographer, as to the Limits of Louisiana, Etc.," *Historical Magazine* 3 (1859): 231–32 [Hereafter HM]. As of yet, no French originals (or copies) have been found to correspond to the four letters translated in English appearing in HM.

23. AN, Marine, 2JJ 60, XVI #54. See Dawson, *L'Atelier Delisle*, 109; and Petto, *When France was King*, 127, n. 22.

24. AN, Marine, 2JJ 60, XVI, #54. Petto, *When France was King*, 127, n. 23.

25. Letter Bobé to Delisle, 8 January 1715, HM, 231.

26. Giraud, *French Louisiana*, vol. 1, 249–55. While it is very likely that Crozat had nothing more than a very narrow commercial policy in mind for Louisiana from which he would benefit greatly, the crown was indecisive in its commitment to establishing a colony of both great economic and strategic importance. There had been an effort to keep Crozat's letters patent private by Nicolas Desmarets, the controller-general of finances, who argued against publicizing the crown's renewed interest in a "company project" in Louisiana.

27. Giraud, *French Louisiana*, vol. 1, 259.

28. Jack Jackson, "The Soupart Map of 1716 and its Influence on Gulf Cartography," *Mapline*, nos. 80–81 (Winter 1996/1997): 10–11.

29. Jackson, "Soupart," 10.

30. See AN, Marine, 2 JJ 56, X, 26; Marcel Giraud, *Histoire de La Louisiane Française*, Vol. 2 (Paris: Presses Universitaires de France, 1958), 12–26; Dawson, *L'Atelier Delisle*, 114–16; Giraud, *French Louisiana*, Vol. 1, 366–69; and Jack Jackson and Robert S. Weddle, "The Oliván Rebolledo-Saint-Denis Maps of Texas, Louisiana, and New Spain, 1715–1717," in *Mapping Texas and the Gulf Coast: The Contributions of Saint-Denis, Oliván, and Le Maire*, ed. Jack Jackson, Robert Weddle, and Winston Deville (College Station: Texas A&M University Press, 1990), 7.

31. Sara Stidstone Gronim, "Geography and Persuasion: Maps in British Colonial New York," *The William and Mary Quarterly* 58, no. 2 (2001): 382.

32. Pierre Pluchon, *Histoire de la Colonisation Française* (Paris: Libraire Arthème Fayard, 1991), 352; and Giraud, *French Louisiana*, vol. 1, 14–15.

33. Cecil Headlam, *Calendar of State Papers, Colonial Series, America and West Indies, March 1720–December 1721* (London: His Majesty's Stationary Office, 1933), 202–7, 26 November 1720, #303.

34. Headlam, *State Papers, 1720/21*, 203.

35. Headlam, *State Papers, 1720/21*, 203.

36. Headlam, *State Papers, 1720/21*, 203.

37. Cecil Headlam, *Calendar of State Papers, Colonial Series, America and West Indies, August 1717-December 1718* (London: Her Majesty's Stationery Office, 1930), 119–20, 7 December 1717, #238.

38. Headlam, *State Papers, 1714–1715*, 232, July 1715, #518.

39. Cecil Headlam, *Calendar of State Papers, Colonial Series, America and West Indies, January 1719 to February 1720* (London: His Majesty's Stationery Office, 1933), 187–88, August 1719, #354
40. Headlam, *State Papers, 1720–1721*, 347–48, 18 June 1721, #534.
41. Dennis Reinhartz, "Herman Moll, Geographer: An Early Eighteenth-Century European View of the American Southwest," in *The Mapping of the American Southwest*, ed. Dennis Reinhartz and Charles C. Colley (College Station: Texas A&M University, 1987), 32–33.
42. Herman Moll, *A New Map of the North Parts of America Claimed by France* (London, 1720), BL, Maps, 69917 (29).
43. Dennis Reinhartz, *The Cartographer and the Literati: Herman Moll and his Intellectual Circle* (New York: Edwin Mellon Press, 1997), 71–89, 114–17. In an effort to compete with the French and the Spanish in the South Pacific, England formed the joint stock company, the British South Sea Company in 1711. Fraudulent activity had artificially increased the value of the stock, but in late 1720 when the "bubble" burst and stock prices fell drastically, a credit collapse and financial crisis ensued.
44. See BL, *Representation of the Lords Commissioners for Trade and Plantations upon the State of His Majesty's Colonies in the Continent of North America*, Add. Ms. 23615, f. 72, 8 September 1721. This document addressing French pretensions in North America includes a manuscript map by Popple. Also see BL, *Hardwicke Papers, Reports and Papers Relating to the Plantations in America, 1721–1736*, Add. Ms. 35907, f. 1 is Popple's 1733 map and f. 2–49 is the same report from Add. Ms. 23615.
45. Gronim, "Geography and Persuasion," 383.
46. K. G. Davies, ed., *Calendar of State Papers, Colonial Series, America and West Indies* vol. 45 (London: H. M. Stationary Office, 1994), 109, 24 May 1739, # 179.
47. Henry Popple, *A Map of the British Empire in America with the French and Spanish Settlements Adjacent Thereto* (London, 1733), BL, Maps, 69917 (4).
48. Popple, *Map of British Empire in America.*
49. Gronim, "Geography and Persuasion," 386.
50. William P. Cumming, *British Maps of Colonial America* (Chicago: University of Chicago Press, 1974), 12.
51. William P. Cumming, *The Southeast in Early Maps* (Chapel Hill: University of North Carolina, 1962), 198–99.
52. See response to the February 1738 queries in K. G. Davies, ed., *Calendar of State Papers, Colonial Series, America and West Indies*, vol. 44 (London: H. M. Stationary Office, 1969), 126–30, 2 June 1738, #268, ii. http://www.british-history.ac.uk/report.aspx?compid=72948.
53. Davies, *State Papers*, vol. 44, 126–30.
54. . Jean Baptiste Bourguignon D'Anville, *Carte de l'Amerique septentrionale dressée pour l'usage particulier de Mgr. le Duc de Chartres . . .* (Paris, 1736), BN, Cartes, GE D 10638.
55. De Vorsey, "'Selling' the Georgia Scheme," 35.
56. De Vorsey, "'Selling' the Georgia Scheme," 38–39.
57. De Vorsey, "'Selling' the Georgia Scheme," 39.
58. De Vorsey, "'Selling' the Georgia Scheme," 38–39.
59. De Vorsey, "'Selling' the Georgia Scheme," 36.
60. Davies, *State Papers*, vol. 45, 74 (2 April 1739): #125.
61. Mart A. Stewart, "William Gerard de Brahm's 1757 Map of South Carolina and Georgia," *Environmental History* 16, no. 3 (2011): 524–25.
62. "Account of British Settlements in America," *Gentleman's Magazine* 26 (1756): 20.
63. Malachy Postlewahyt, *Britain's Commercial Interests Explained and Improved*, vol. 1 (London, 1757), 459.
64. Some of the material for this section is drawn from Petto, *When France was King*, 106–13.
65. In aftermath of the Treaty of Utrecht (1713), the intendant of Canada, Michel Bégon de la Picardière, and the governor-general, Philippe de Rigaud, marquis de Vaudreuil, responded to a royal report concerning the boundaries of Acadia. See Vaudreuil letter concerning limits,

http://collectionscanada.gc.ca/pam_archives/index.php?fuseaction=genitem.displayItem&lang=eng&rec_nbr=2487097&rec_nbr_list=2313741,2313745,2487097,2487440,2318933,2487474,99938,2313790,2313763,2487055 (accessed 17 June 2013).

66. Petto, *When France was King*, 107.

67. Jacques Nicolas Bellin, "Remarques de M. Bellin, ingenieur de la Marine, sur les Cartes et Plans," in *Histoire et Description de la Nouvelle France*, ed. Père Charlevoix (Paris: Giffart, 1744), vol. 3: xii.

68. Bellin, "Remarques," vol. 3: vi. "On sera peut-être surprise de ne pas trouver des sondes sur mes plans; c'est-à-dire, la quantité de brasses, ou de pieds d'eau; je sçais que ces détails sont extrêmement utiles, & il m'auroit été facile de les remplir avec exactitude: mais des raisons particulieres, qui n'ont rien de commun avec la Géographie, m'en ont empêché. A l'égard des plans des Ports, qui n'appartiennent pas à la France, j'y ai mis des sondes." See also Mireille Pastoureau, "Jacques Nicolas Bellin, French Hydrographer, and the Royal Society in the Eighteenth Century," *Yale University Library Gazette* 67 (1993): 68; and Mary Sponberg Pedley, *The Commerce of Cartography: Making and Marketing Maps in Eighteenth-Century France and England* (Chicago: University of Chicago Press, 2005), 168–69.

69. "Remarks by M. Bellin, in relation to his Maps drawn for P. Charlevoix's *History of New France*," *Gentleman's Magazine* 16 (1746): 71–75. "The readers will perhaps wonder why I have not in these plans mark'd the soundings, the fathoms or depths of water; a point of great use, and which I could easily have ascertain'd, but particular reasons, foreign to geography, prevented me. With regard to such ports as do not belong to France, I have been very exact in this particular" (73).

70. Thomas Jefferys, *A New Chart of the coast of New England, Nova Scotia, New France or Canada . . .* in *The Gentleman's Magazine* 16 (1746): 72.

71. The government attitude that attention to the continental war (War of Austrian Succession, 1740-48), would ultimately take care of the colonies—"*il faut conquérir l'Amérique en Allemagne*"—allowed the English navy, gaining in strength, to interrupt French trade with America. As quoted in James Pritchard, *Anatomy of a Naval Disaster: The 1746 French Expedition to North America* (Montreal: McGill-Queen's University Press, 1995), 15.

72. See Petto, *When France was King*, 129, n. 49.

73. James Pritchard, *Louis XV's Navy 1748-1762* (Montreal: McGill-Queen's University Press, 1987), 7, 27.

74. Roland-Michel Barrin marquis de La Galissonière, *Mémoire sur les colonies de la France dans l'Amérique Septentrionale, Décembre 1750*, http://collectionscanada.gc.ca/pam_archives/index.php?fuseaction=genitem.displayItem&lang=eng&rec_nbr=3072107&rec_nbr_list=3071693,3072107,3071692,3072443,3071703,100365,98727 (accessed 16 October 2013).

75. McClellan and Regourd, *Colonial Machine*, 182.

76. Joseph-Bernard, marquis de Chabert, *Voyage fait par ordre du roi en 1750 et 1751 dans l'Amerique septentrionale, pour rectifier les cartes des côtes de l'Acadie, de l'Isle-Royale et de l'isle de Terre-Neuve* (Imprimerie Royale, Paris, 1753), 7. "Elle me prescrivoit de leur montrer mon instruction, présumant qu'ils ne s'opposeroient pas à un travail qui avoit pour objet la sûreté commune de tous les Navigateurs."

77. "Account of the Best Foreign Books Published in 1754," *Gentleman's Magazine* 24 (1754): 600.

78. "A Summary of all the positions of Places determined by M. Chabert," *Gentleman's Magazine* 25 (1755): 128.

79. John A. Schutz, *William Shirley, King's Governor of Massachusetts* (Chapel Hill: University of North Carolina, 1961), 153.

80. Schutz, *William Shirley*, 159.

81. Schutz, *William Shirley*, 166. Shirley pressed forward to present his highly detailed reports and retorts to both the French commissaries and the London officials, but by April 1752 he was dismissed from his post. With the duke of Bedford out of office (summer 1751), Newcastle's faction in power, and tension with his co-commissioner, William Mildmay, who "accused [Shirley] of deadlocking negotiating for two years," Shirley's exacting inflexibility (and perhaps his dalliance and marriage to his Parisian landlord's daughter) led to his dismissal.

Mildmay forged ahead seeking a "policy of containment, for drawing boundaries of empire in America and establishing neutral zones between the empires." The Boundary Commission failed. "In the end he failed to win French cooperation, and events in America proved that the rivalry for empire could not be avoided." Also see Theodore Calvin Pease, *Anglo-French Boundary Disputes in the West, 1749–1763* (Springfield: University of Illinois, 1936).

82. "Limits of Nova Scotia," *Gentleman's Magazine*, 20 (1750), 82.

83. "A Further Account of Nova Scotia," *Gentleman's Magazine*, 20 (1750): 264. My emphasis in the quote.

84. "The Importance of Nova Scotia," *Gentleman's Magazine*, 18 (1748): 29.

85. See Mary Sponberg Pedley, "Maps Wars: The Role of Maps in the Nova Scotia/Acadia Boundary Dispute of 1750," *Imago Mundi* 50 (1998): 96–104.

86. William Shirley and William Mildmay, "A Memorial Concerning the Limits of Nova Scotia or Acadia, . . . January 11, 1751" in *The Memorials of the English and French Commissaries Concerning the Limits of Nova Scotia or Acadia*, vol. I (London: 1755), 71.

87. William Mildmay and Ruvigny de Cosne, "Reply of the English Commissaries . . . to their [French] Memorial of the 4th of October 1751, concerning Nova Scotia or Acadia," in *The Memorials of the English and French Commissaries Concerning the Limits of Nova Scotia or Acadia*, vol. I (London, 1755), 263–65.

88. Mildmay and de Cosne, "Reply of English Commissaries," vol. I, 263–65.

89. Shirley and Mildmay, "Memorial January 11, 1751," vol. I, 73.

90. Roland-Michel Barrin, marquis de La Galissonière and Etienne de Silhouette, "Mémoires des Commissaires du Roi, du 4 Octobre 1751 . . . concernant l'Acadie," in *Mémoires des Commissaires du Roi et de ceux de sa Majesté Britannique, sur les Posessions & les Droits respectifs des deux Couronnes en Amérique*, vol. I (Paris: Imprimerie Royale, 1755), 124. "On conviendra, avec les Commissaires de Sa Majesté Britannique, que leur [geographes] autorité ne doit point être décisive. Ils sont plus occupés de donner un air de système & de vérité à leurs cartes, ainsi qu'une apparence de science & de recherche, qu'à fixer les droits des Princes, & les véritables llimites des pays."

91. Shirley and Mildmay, "Reply of English Commissaries," vol. I, 271.

92. Shirley and Mildmay, "Reply of English Commissaries," vol. I, 281, 283.

93. Pedley, "Maps Wars," 97.

94. Pedley, "Maps Wars," 97.

95. Jacques Nicolas Bellin, *Carte de la Partie Orientale de la Nouvelle France ou de Canada dediée à Monseigneur le comte de Maurepas Ministre et Secretaire d'Etat Commandeur des orders du Roy*, in *Histoire et Description de la Nouvelle France*, ed. Père Charlevoix (Paris: Giffart, 1744), vol. I, plate #8. Henry Popple, *A Map of the British Empire in America with the French and Spanish Settlements adjacent thereto . . .* (London, 1733).

96. Shirley and Mildmay, "Memorial January 11, 1751," vol. I, 69–71. "The next is the Chart of the Sieur *Bellin*, published in 1744, in which the Limits of *Acadia, under the Name of Nova Scotia*, are marked to extend as far Westward as the River *Penobscot*, or *Pentagoet*, to which River likewise the Eastern Bounds of *New-England* are there made to extend; and the Bounds of *Acadia*, under the Name of *Nova Scotia*, are carried across the Continent many Leagues Northward of the *Isthmus*, to the Gulph of *St. Laurence*, as high as the opposite Northernmost Point of the Island of St. *John*, and no Part of *New France* or *Canada*, is mark'd on the Souuthern side of the River St. *Laurence*."

97. La Galissonière and Silhouette, "Mémoires 4 Octobre 1751," vol. I, 122. "Le sieur Bellin a été visiblement induit en erreur par les cartes & les idées Angloises, en supposant qu'il existoit une nouvelle Ecosse réelle, distincte & indépendante de l'Acadie . . . mais au surplus, il a restreint l'Acadie dans la péninsula; & en ce point, qui est le point essential & capital, le seul auquel se réduit l'état de la question, la carte qu'il a faite, & que produisent les Commissaires Anglois, est directement contraire à leurs prétentions."

98. Louis De Vorsey, *The Georgia-South Carolina Boundary: A Problem in Historical Geography* (Athens: University of Georgia Press, 1982), 24.

99. Henry Popple, *A Map of the British Empire in America* (London, 1733), BL, Maps 69917 (4).

100. Mildmay and de Cosne, "Reply of English Commissaries," vol. I, 277.

101. Mildmay and de Cosne, “Reply of English Commissaries,” vol. I, 279.
102. La Galissonière and Silhouette, “Mémoires 4 Octobre 1751,” vol. I, 177. “Les Commissaires Anglois, pour determiner des limites anciennes, on eu recours à des cartes modernes; mais la plûpart même des cartes modernes, & toutes les anciennes, restreignent l’Acadie dans la peninsule, ou dans une partie seulement.”
103. Mildmay and de Cosne, “Reply of the English Commissaries,” vol. I, 283.
104. For the presentation on the Vaugondy map and the dispute, see Mary Sponberg Pedley, *Bel et Utile*: The Work of the Robert de Vaugondy Family of Mapmakers (Tring: Map Collector Publications, 1992), 69–74. In 1755, Vaugondy published *Partie de l’Amérique Septentrionale qui comprend la Nouvelle France ou le Canada*, in which he continued to present Acadia as a mere strip of land on the peninsula and confined New England (and thus English lands) to south of the Penobscot River. See BN, Cartes, GE DD 2987 (8577).
105. Pedley, *Bel et Utile*, 72.
106. *Mercure de France* 4 (Juillet 1753), 165–66. “Nouvelle Carte du Canada, dédiée & présentée à M. le Comte d’Argenson, Ministre & Secrétaire d’Etat; par M. Robert de Vaugondy fils, Géographe ordinaire du Roi, en Juin 1753. L’on trouve dans cette Carte un detail que l’on ne peut avoir dans aucune de celles qui ont paru jusqu’à present. L’Auteur y a distingue les possessions Françoises & Angloises exactement, sur tout pour l’Acadie, d’après les Mémoires qui lui avoient été communiqués il y a un an par le Ministere, pour corriger l’article de cette Prequ’isle dans le Dictionnaire de Trévoux. Des Ouvrages si bien executes contribueront à soutenir la reputation qu’ont mérités à l’Auteur les grands globes que nous avons announces dans quelques uns de nos Mercures.”
107. *Mercure de France* 4 (Septembre 1753), 155–56. Lettre à l’Auteur du Mercure, “Je vous prie, Monsieur, de vouloir bien, par votre plus prochain Journal, désabuser le Public sur l’impression qu’a pû faire l’annonce de ma Carte du Canada, dans votre Mercure de Juillet. Mon intention, Monsieur, n’a pas été de faire entendre que j’eusse reçu aucun mémoire du Ministere. Quant à ce qui constitue le travail de ma Carte, je me suis fait un devoir de rassembler sur cette partie ce que j’ai connu de meilleur, tant dans les morceaux déja publiés que dans les manuscrits dont j’ai eu connoissance. Je me garderai bien de m’arroger pour cela aucune supériorité sur ceux qui ont couru la même carrier que moi, & dont je reconnois tout le mérite. Le Public a leurs ouvrages entre ses mains, ainsi que l e mien, c’est à lui seul à en juger. J’ai l’honneur d’être, &c. Robert de Vaugondy.”
108. John Green, *The Conduct of the French with regard to Nova Scotia* (London, 1754), 2.
109. Green, *Conduct of French*, 9. “On this occasion, it may be observed as a common rule, that they who confess against themselves, are more to be believed than those who deny for themselves. But, in thus opposing one French authority to another, I do not quote one of their ordinary geographers: for Mr. Bellin is hydrographer to the marine, as well as censor royal; and his contradicting the author whom he was employ’d to illustrate, gives a double force to his authority.—If therefore, in a subsequent map of the same country, he hath omitted the name of Nova Scotia, it was not, as may be presumed, in consequence of being better informed, but because he was otherwise directed or inclined.”
110. Green, *Conduct of the French*, 46.
111. Green, *Conduct of the French*, 50–51.
112. John Green, *A New Map of Nova Scotia and Cape Britain, with the adjacent parts of New England and Canada, composed from a great number of actual surveys; and other materials regulated by many new astronomical observations of the longitude as well as latitude with an explanation* London, 1755, BL, Maps 184.i.3.(8.); John Green, *Explanation for the New Map of Nova Scotia* . . . (London, 1755), BL, Maps, K. Top. 119.11.b.
113. Green, *Explanation for New Map*, 13, 16.
114. Green, *Explanation for New Map*, 13. “This great number of monstrous errors of every kind, in the figure and position, as well as longitude and latitude of places, is, methinks, a reproach both to navigators and geographers. To navigators, for neglecting to make exact observations at all ports and headlands wherever they come; altho’ they see the wretched condition of the charts. Which are every moment before their eyes. To geographers, because, notwithstanding their materials of any worth are so few and imperfect, they industriously conceal them from one another, which hinders the advancement of the sciences: whereas if they

would communicate them, the maps resulting from their several labours would be more accurate and complete, if not perfect; nor would they differ so enormously, if they did not altogether agree. It is with a view to bring about such a communication of materials, that I have been so particular in specifying those which I made use of."

115. Green, *Explanation for New Map*, 17.

116. Jacques Nicolas Bellin, *Carte de l'Amerique Septentrionale* (Paris, 1755), BN, Cartes, GE DD 2987 (8535B); and *Remarques sur la Carte de l'Amérique Septentrionale* (Paris: Didot, 1755).

117. Bellin, *Remarques*, 32–33.

118. Bellin, *Remarques*, 5.

119. Bellin, *Remarques*, 5–6.

120. Matthew Edney, "John Mitchell's Map of North America (1755): A Study of the Use and Publication of Official Maps in Eighteenth-Century Britain," *Imago Mundi* 60, no. 1 (2008): 63–85, here, 71.

121. Edney, "Mitchell's Map," 72; K. H. Ledward, ed. *Journals of the Board of Trade and Plantations, Volume 9: January 1750–December 1753* (London: H. M. Stationary Office, 1932), 85–91, 18 July 1750, "Plantations General." http://www.british-history.ac.uk/report.aspx?compid=81833 (accessed: 29 June 2014).

122. John Mitchell, *A Map of the British and French Dominions in North America . . .* (London, 1755), BL, Maps CC.5.a.270.

123. Full quotation: "The map of New England & Nova Scotia requires a further consideration as we find them very erroneously laid down in all our maps & charts, especially our many new maps, copyed from a new map of Nova Scotia copyed from Popple and d'Anville and those errors are maintained by arguments & pretended authorities which seem to have confirmed them. The only authority they have for all this is a feigned survey by a pretended surveyor general Blackmore in 1711–12 who appears by his journals to have been lieutenant of the Dragon Man of War 1711 and made a rude Draught of this coast (as well as he remembered it perhaps) in 1715, with a petition to the Board of Trade to enable him to survey it at that time which he never did as we can learn. But this Draught falling into the hands of some workmen, Mr. Moll published it as an actual survey 'made by her majesties especial command' from which this coast has been thus erroneously laid down ever since. . . . [in concluding the long blurb] In short we do not find a single spot hardly justly laid down in these our New Maps of Nova Scotia, altho they were rightly laid down formerly by de L'ilse & others ___. We are so far from improving then in the Geography of America that we see it made worse & worse for want of certain observations, which we have endeavoured to collect & thus to represent in one view."

124. Edney, "Mitchell's Map," 75.

125. Edney, "Mitchell's Map," 78.

126. Edney, "Mitchell's Map," 79.

127. . Thomas Jefferys, *Remarks on the French Memorials concerning the Limits of Acadia; printed at the royal printing house of Paris, and distributed by the French ministers at all the foreign courts of Europe with two maps exhibiting the limits one according to the system of the French, as inserted in said memorial; the other conformable to the English rights; as supported by the authority of treaties, continual grants of the French kings, and express passages of the best French authors to which is added an answer to the summary discussion, &c.* (London, 1756), BL, Maps 8176.aa.39.

128. Jefferys, *Remarks on the French Memorials*, 37. BL, Maps 8176.aa.39.

129. Pedley, *Commerce*, 120 "After their loss of Canada at the end of the Seven Years' War, the French were also reliant on the British map trade for public information about the coasts, interior, and infrastructure of eastern North America and the Atlantic coast.

130. Stephen J. Hornsby, *Surveyors of Empire: Samuel Holland, J. F. W. Des Barres, and the Making of the Atlantic Neptune* (Montreal: McGill-Queen's University Press, 2011), 4.

131. Hornsby, *Surveyors of Empire*, 5–6.

Chapter Five

Charting the Seas of the East Indies

Commercial Opportunism versus Royal Approbation

> The French have been at pains, to improve their Navigation and their Charts; Those of the Indian Seas, by Monsieur D'Après de Mannevillette, a Captain in their Service, exceed every thing of the kind in Europe. It is a pity they are not translated into English, for the benefit of our Navigators. Those who understand the language, and have seen them, must have a despicable opinion, of our Indian Pilot, with which Messrs. Mount and Page have long imposed on, and picked the pockets of our Countrymen, and which are only fit for the Grocers and Chandlers Shops, or posterior uses. —Charles Noble, 1755

By the seventeenth century, England, France, and the United Provinces would come to challenge the trading dominance of Portugal in the East Indies. The United Netherlands Chartered East India Company (hereafter, VOC) would succeed initially followed closely behind by the English and, later, by the French trading companies.[1] While England would be more commercially successful in challenging the Dutch, France—despite inconsistent governmental attention—continued a trading (and diplomatic) presence even upon its severe reduction in the region after the Seven Years' War. Initially and into the eighteenth century, both England and France continued to rely on Dutch charts and traditional sailing directions. By the mid-eighteenth century and ironically just a few decades before France's presence in the area was reduced by British dominance, a captain from the Compagnie des Indes Orientale (Company of the East Indies) produced an atlas of maps and sailing directions better than anything then available.

For the first time since Pierre Garcie's *Routier*, a pilot produced a work for pilots; while much of the hydrographic work of the eighteenth century would still come from the hand of bureau map and chart makers, there was a continual appeal from their offices as well as from leading navigators that those in the field with direct experience play a role in improving charts. These works became part of a "geographical archive"[2] inspired initially by seventeenth-century Baconian science and then further developed by the Enlightenment epistemology that joined empiricism and rationalism in this hydrographic sphere for the surety of navigation. The challenge, however, was for this ideology both to propel the work of navigators in the field forward under the guiding principles of making navigation safer and to uphold in the bureau the attention of the geographer/hydrographer of the study to a reasoned argument to reconcile sources similarly for that surety. In the absence of nineteenth-century marine surveying expeditions, the field and the study needed to act in coordination. Necessary for that coordination and the fulfillment of the Enlightenment ideology was the presence of an institutional framework which in France, according to James McClellan and François Regourd, was a multi-institutional structure, that is, the Colonial Machine which would not begin to appear in Britain until the late eighteenth century with the establishment of the Hydrographic Office (1779) of the British East India Company and the Admiralty's Hydrographic Office in 1795.[3] In contrast, a challenge to the fulfillment of this ideology was the role of the commercial chart trade which both helped to disseminate information but also potentially stifled the production of up dated work. The London chart trade in which pilots could sell their charts to publishers continued to be commercially driven into the last decades of the eighteenth century, and many a pilot would rather a French chart than a British one.

The focus of this chapter is on the role of charting in the economic and political ventures supported by these trading companies and their respective governments. The story of charting the East Indies has three parts: the formation and early actions of the commercial trading companies in the seventeenth century and their motivation (or support) for charting into the eighteenth century; the increased presence of British and French shipping and efforts for more charting of the region in the mid-eighteenth century and the scientific context of this charting (in particular, the direction by those institutions belonging to the French Colonial Machine); and the tension between commercial endeavors, exploratory expeditions, and war/diplomacy in the context of charting in the second half of the eighteenth century.

SEVENTEENTH-CENTURY ENGLISH AND FRENCH TRADING COMPANIES FOR THE EAST INDIES

England first came to challenge the Iberians and the Dutch in the Spice Islands in the early seventeenth century.[4] London merchants and elite investors had organized into a corporation and sought recognition from Elizabeth's Privy Council and the grant of a monopoly on English Asian trade. In late 1600 the English East India Company was chartered and a few months later, planned its first expedition which was expressly to be for commercial purposes and, therefore, instructed to avoid hostile encounters with, in particular, the Portuguese.[5] Richard Hakluyt provided the Company with information on the commodities available and the major trading centers, as well as three maps on the Spice Islands most likely drawn from the Portuguese originals found on board the *Madre de Deus* taken off the Azores in 1592.[6] Over the course of the first half century, though not without challenges from other English entrepreneurial groups and periods of hostility with both native and other European powers (i.e., the Dutch and the Iberians, mostly Portuguese), the Company set up nearly two dozen factories (e.g., Surat, Masulipatam, Fort St. George, Bombay, and Bantam) in the eastern regions. They sometimes cooperated (briefly) with the Dutch against the Portuguese traders, and gradually established relationships with local rulers in the region. By the end of the Thirty Years' War (1648), the United Provinces (or Dutch Republic) was free from Spanish control, and the Dutch became a more dangerous competitor to the English leading to the Anglo-Dutch Wars in the second half of the seventeenth century. While England countered with greater efforts in the region to challenge Dutch power and influence, it had to survive the economic downturn in the 1640s and 1650s due to the Civil War at home.[7] Emerging from the period of the Interregnum was an East India Company that had survived its competitors (at home) and regained the privileges accorded it under Elizabeth.

Within a couple decades, however, there were voices raised in opposition to the "oligarchic" Company, its restrictive participation, and its monopoly on Asian trade.[8] In the post-Glorious Revolution years, the opposition won out and a New East India Company was established in 1698, and both companies operated in parallel for about a decade. Soon the reality of how entrenched the old Company was led to a merger of the two into the United East India Company, 1709 (hereafter, EIC). At the beginning of the eighteenth century it was clear that Asian trade was quite valuable to England and its domestic port of call, London. At twice the value of trade from North America and comparable to that from the West Indies, this was a destination worth more attention.[9]

By the mid-seventeenth century, the financing mechanism for Company voyages had already changed from investing individual voyages to a more permanent joint stock company which prompted the need to plan for annual voyages and to support an infrastructure for such endeavors, including maritime knowledge.[10] Early in the Restoration period, Charles II, having received Bombay from Catherine of Braganza's dowry, transferred it to the Company (1668) and accorded the Company sovereign powers in the region to declare war, negotiate peace, mint coins, name their governor and to impose local jurisdictional control. The Company was growing into a powerful force in the region and a useful player in public finance at home.[11] The demands, however, of carrying on annual voyages and expanding trade further east put greater and greater stress on the need to accumulate knowledge, that is, currents, coastlines, and sailing directions, for navigational purposes.[12] At this point in the seventeenth century while England was gaining a foothold in East Indies trade, France was only just getting started.

In terms of trade, the French were *late to the party*, in effect, having only sporadic interest in the East Indies up to the mid-seventeenth century.[13] In the 1660s Colbert initiated a more stable and potentially successful venture, the Compagnie des Indes Orientale (hereafter, CIO), which unlike the London Company was *born* of the Crown's ministry.[14] As one of the sponsoring institutions of the French Colonial Machine, it was "'tied tightly to the state.'"[15] Under the direction of Colbert, Louis' finance minister and architect of the projects of *gloire*, François Charpentier (1620–1702) had published a pamphlet to excite interest in an East Indies trading company.[16] Charpentier reminded his readers that the war that had occupied France in the previous century and the unrest of previous decades had ended, and it was time for France (with so many great ports, vessels, experienced seamen, and brave soldiers) to travel to those places in which its neighbors had already established themselves. He chided the French for not making use of the work that others had done in setting off to these distant lands. With nationalistic fervor, he wrote that now that God had returned tranquility to France, and it enjoyed a glorious peace under Louis XIV, whose wise governance was an object of admiration and fear from all of Europe, the French must take their place to establish a glorious commerce of an inestimable utility already enjoyed by its neighbors.[17] To provide experienced leadership for the CIO, Colbert "drafted" François Caron (1600–1673), a French Huguenot who had worked his way through the VOC organization to Director-General at Batavia, the center of Dutch East India and the location of its *in situ* hydrographic office. Caron would become the Director-General for the CIO from 1667–1673. With its fifty-year monopoly soon the French company had set itself up in Surat as the hub of operations establishing their first two trading posts at Pondicherry and Chandernagor. Similar to the EIC, the volume and value of trade from Asia far outweighed that coming from North America, and France intended to challenge the Dutch and the English for a greater share of the Asian trade.[18]

EARLY CHARTING BY ENGLAND AND FRANCE

Prior to the early eighteenth century, there was no official institution responsible for chart making in either England or France.[19] In London, pilots might turn in their ships' journal to the Admiralty office or to the India House, but there was no systematic procedure or plan for gathering such information. Pilots returning from their eastern voyages might give information to chart makers whose business establishments were along the River Thames and ask them to produce useful charts for upcoming travels. Similarly, pilots setting out for these voyages might seek out these establishments for maritime charts of the region.[20] Pilots, however, may not have been pleased with the resulting charts as Anthony Hippon, Captain of the *Globe*, wrote in May 1612 to the Company and complained of the inaccuracy of a chart by one of these chart makers, John Daniels (d. 1649), covering Cape Comorin and the Western part of Ceylon. "I would advise your worships that charge is given unto all such mariners as ar [sic] entertained in your service not to buy any orf [sic] those erroneous maps."[21] These were manuscript charts relying on the quality of information provided and the expertise of the chart maker. Recall that John Seller's project to produce a sea atlas (in several books) covering the world (see chapter 3) ended up relying on old Dutch plates, and the book covering the East Indies was left incomplete until John Thornton revised the work and published it in 1703 (see below). Similarly, in France pilots returning from the East Indies might submit their journals to the CIO office in Lorient or to the office of the naval minister. Aside from Colbert's official statement (1681) that pilots turn in their ships' journals to the clerk of the admiralty so that important information could be gathered by the naval ministry, there was no plan beyond the accumulation of documents.[22] There was (as yet) no institutional means in either country to systematically gather, critically examine, synthesize and re-analyze, draw, engrave, and publish charts with instructions for gathering specific information. The Enlightenment "geographical archive" was still a disparate miscellany of travel itineraries and journals, and geographic, hydrographic, and navigational information.

Along the River Thames, chart makers served all pilots arriving from near and far destinations and while as seen above not all navigators were pleased with the resulting charts, the Thames School chart makers acquired a good reputation for their works. John Thornton (1641–1708), a member of this chart-making group, became the hydrographer of the East India Company and Hudson Bay Company. He produced several manuscript maps based not only on new information being provided by these pilots, but also by England's great competitor in the region, the Dutch, whose published works from the seventeenth century included charts from the publishing houses of Blaeu, Van Keulen, and Goos. With a fellow member of the Thames School, Joel Gascoyne (1650–c. 1704), he produced a large oceanic chart of the Indian Ocean (c. 1684) and dedicated it to Henry Somerset (1629–1700), Duke of Beaufort and a member of the Committee of the East India Company (1684–1690).[23] Unlike much of Thornton's work, this chart was engraved and printed. Perhaps to accompany his grand chart, Thornton produced two smaller charts one covering the island of Bombay along the western coast of India and the other the Ganges River and Bay of Bengal along the northeastern coast of India (see below). Both charts carried the coat of arms of the Company and a dedication to the Governor, Deputy Governor, and Committee of the East India Company.

The Governor at the time, Sir Josiah Child (1630–1699), was a staunch supporter of mercantilist policies and an advocate for a strong show of power against the Dutch by establishing fortified settlements independent of local rulers. "He proposed to build up Madras and Bombay into such [fortified] settlements and to take a suitable base in Bengal by force."[24] While the 1686 expedition that he sent to challenge Mughal power was unsuccessful, Thornton's charts may have helped give the Committee a view from a distance with his oceanic chart for the Duke and his two more focused charts. Thornton's chart on the Ganges River (1685), *A Mapp of the Greate River Ganges as it Emptieth it Selfe into the Bay of Bengala*, was by its title based on new information from Company agents and never before made public.[25] Thornton's pilotage chart of Bombay, *A New Map of the Island of Bombay and Sallset*, included soundings, coastal hazards, a compass rose, the latitude of Bombay Island (19 degrees), notations for towns, fortresses, a church, and the salt ponds and the depiction of a Company ship headed into port.[26]

For the pilot of that Company ship and so many like him, Thornton, in 1703, revised and expanded John Seller's *The English Pilot, Third Book on Oriental Navigation*. This work provided the first "systematic" presentation of knowledge for sailing from England to the East Indies and navigating in the eastern waters.[27] Thornton's chart making had perhaps benefited from the Glorious Revolution as with William of Orange as king of England, access to Dutch charts and maps had improved.[28] For example, in 1703 upon the capture of an English East Indiaman, the *Canterbury*, the French came into possession of 17 manuscript charts dated between 1699 and 1701 and all but one signed by Thornton. The information on these charts is thought to have come from Dutch sources.[29] The national flags (e.g., Dutch, French, English) present on the manuscript charts are absent on the printed charts perhaps so as not to memorialize the presence of a commercial competitor one intended to replace.[30]

As much of navigation was still reliant on sailing directions, Thornton included information for "shaping the course" as the vessel sailed from one port to another at various times of the year according to the monsoonal winds and currents. In addition, he supplied latitude and longitude values for principal locations from Land's End in England to the Cape of Good Hope, coastal views and profiles, as well as the variation of the compass from observations made in 1700 and 1701. In a new third section with information on compass variation, he extended the voyagers' knowledge by covering the "sea coasts, capes, head-lands, soundings, sands, shoals, rocks, dangers, islands, bays, roads, harbours, and ports from Cape Good Hope to the eastern seas."[31] According to Thornton, such a work drawn from the drafts and descriptions of English and Dutch navigators had not yet been published, and as a worthy servant he dedicated the entire work to the "mannagers [sic] for the United Trade to the East-Indies."[32] The work opened with a world map in Mercator projection (i.e., according to Edward Wright) depicting the passages of Edmond Halley (1698) to South America and the Southern Ocean and William Dampier (1687) to Australia and the Eastern Islands. The second map was drawn from Halley's chart of the English Channel (see chapter 3), the starting and ending point for English voyages. In addition to the earlier 1685 charts covering Bombay and the Ganges River, there were several oceanic and pilotage charts drawn from those manuscript charts—found on the *Canterbury*—which often included not only coastal hazards and soundings, but also some coastal views. It was, therefore, at the time among the best source of hydrographic and navigational information.

Not surprisingly a few decades later Thornton's *Pilot* joined the sailing directions from Captain H. Cornwall's *Observations upon Several Voyages to India* as the major sources of information for British navigators into the second half of the eighteenth century.[33] By the second decade of the eighteenth century, Thornton's work continued to be published but now by the commercial establishment of Mount and Page, who bought the plates from Thornton's son, Samuel (d. 1715). This *English Pilot* continued to be reissued with few if any changes to the charts (or sailing directions) up to the last edition in 1761.[34] There had been no impetus to update the information over the course of the first half of the eighteenth century. Pilots simply added notes to their rutters and charts and passed their knowledge from experience on to the trusted members of the brotherhood of the sea. In Britain, chart making was too reliant on commercial establishments, and there was (as yet) little motivation on the part of businesses such as Mount and Page to assume any financial expense to improve charts which would reduce profits.

Over the same period of time from the 1680s to 1720, there was not much progress either in chart making by the French for this region. In fact they continued to rely on the works of Pieter Goos (see chapter 3) and the publications by the Van Keulens as well as Thornton's *English Pilot*. Philippe Haudrère has written that the activities of the CIO during this period were much more limited than that of the English and would only begin to change in the second quarter of the eighteenth century.[35] For sailing directions pilots had the seventeenth-century work by F. Dassié, who published them with a work on naval architecture which he had dedicated to Monseigneur Louis Victor le Duc de Vivonne et de Mortemart (1636–1688), marshall of France, general of the galley fleet and a member of the powerful Rochechouart family.[36] Patronage from this general, the brother of Madame de Montespan, mistress of Louis XIV, gave Dassié an influential and interested audience, as he sought to "reduce into Art, as methodically as he could, a Science so necessary and useful to the State, to render it familiar, and to quicken those that are knowing in the Mathematicks and in Naval Architecture, to enquire after infallible ways of making Ships sail better."[37]

As was typical at this time, his sailing directions were drawn from previous works—"extracted out of the most modern and best authors"—from the Portuguese rutter of Aleixo da Mota (c. 1621) and the Dutch work by Jan Huygen van Linschoten (c. 1595) who had worked for the Portuguese in India.[38] Reflecting the lack of reliable charts, Dassié warned that one should not put too much "faith" in maps as they do not truly show where the shoals and islands are. He suggested, however, that it was necessary to pay attention to various markers and make note of them.[39] His warning captured the sentiment of various pilots who would rather put their faith in their sailing directions and their own experience than in any chart. Moreover it is a reminder that pilots stayed true to the route they travelled last and unless knocked off course, continued on the same route time and time again. The desire to open new markets in the eighteenth century, however, would require a search for new information some of which was compiled into the form of a chart.

FIRST HALF OF THE EIGHTEENTH CENTURY

In the first quarter of the eighteenth century, the volume and value of trade by the EIC would surpass that of the Dutch VOC, while the French Company's trade, still expanding, was only half the size and value of the British Company.[40] Both the British and French trade were dominated by that with India supplemented by inter-Asian or "country trade" provided by locals and entrepreneurial Europeans, but as Britain and France vied for a greater economic foothold in the East and more annual voyages—anywhere from a dozen to two dozen—were taking place, company pilots had to pay increasing attention to the routes travelled to make the best use of time, to avoid hazards, and to challenge their competitors.[41] As one set out from an Atlantic (or Channel port) and travelled southwestward following the current, at the Cape Verde Islands the vessels had to turn southwards to avoid following the current to the West Indies. It was, however, also important to pay attention to one's southward progress and, at say Trinidade, within sight of the coast of Brazil, head south or southeast, again following the trade winds in the South Atlantic.[42] Once at the Cape of Good Hope one would make another set of decisions upon entering the Indian Ocean such as those regarding which route to sail to India or China, where to seek provisions including fresh water, when to depart based on the time of year for the prevailing winds (the monsoons), and, of course, the currents.[43] While charts would become increasingly important for navigating this region, sailing directions and coastal profiles remained a fundamental element of pilotage.

With the decline of the Mughal Empire, in the first half of the eighteenth century, both Britain and France sought treaties with regional rulers that would give them favorable trading status over their competitors. Under Pierre Christophe Lenoir (1683–1743), the Governor at Pondicherry, France hoped to work in peace and good relations with the British, but by 1735 under the governorship of Pierre Benoît Dumas (1735–1741) and then Joseph-François Dupleix (1741–1754), the Company's strategy took on not only a more commercially aggressive role but also, regionally, a more politically interventionist role. With Britain and France in the midst of the War of Austrian Succession (1740–1748) on the continent and more conflicts within the French and Indian Wars in their colonial interests in North America, it is not surprising that these European clashes exacerbated tensions in their Indian trading ventures. In 1744, France, in response to British hostilities against French shipping, attacked and took control over Madras (1746–1748) where the illustrious (but short) career of Robert Clive (1725–1774), Britain's Clive of India, began. Moreover, as Britain and France entered the Seven Years' War (1756–1763) both sought new commercial opportunities beyond the subcontinent of India prompting the necessity to find secure shipping routes and to produce new charts of the China Seas in order to address the challenges of threading a vessel through waters decorated with islands and unknown hazards.

And yet despite the volume and expansion of maritime activity by the EIC in the first half of the eighteenth century and the requirements that pilots deposit their journals at the India House at the close of their voyages, there was little effort to provide a new or up-dated systematic presentation of the region in charts. In fact, even though in the eighteenth century pilots were expected to carry charts and instruments and their skills were evaluated by master pilots, neither the EIC nor the government provided the support structure necessary for navigational means and relied on the pilots' mastery which in unfamiliar waters could be quite dangerous.[44] Well into the eighteenth century "commercial opportunism" still directed London chart production for the East Indies.[45] Across the Channel the reported motivation for the domestic production of charts was similarly the surety of navigation and independence from Dutch publications, but new stimuli emerged with the establishment, in 1720, of the navy's hydrographic office, the Dépôt, as well as a renewed CIO initially under the direction of John Law in 1719 and then independent from 1723.[46] These institutions as part of the state's Colonial Machine allowed the French not only to free themselves from Dutch printed charts but to take the lead in map and chart production centered at Paris.

The activities of the Dépôt matched the State's intention to challenge Britain (and the Dutch) on the open seas, and led its hydrographer, Bellin, to produce (by order of the minister of the navy) a series of charts covering these waterways for the royal vessels.[47] For example, in 1740, Bellin produced a map of the eastern ocean for the service of the royal vessels and by the order of comte de Maurepas.[48] On the map he noted that starred locations had had their latitude and longitude determined by astronomical observations; locations with a cross had had their latitude and longitude determined by the height of the pole star; and those places with a double cross had had their latitude observed at sea by several good pilots. He was, in effect, affirming for the reader that empirical data had been used to construct the map. Furthermore, in these remarks he reported that since his chart was "extremely different" than those that had appeared earlier in print, he would accompany the work with a report on the reason for the changes. In that report, he, as usual, wrote about the necessity to improve charts and compared his new work with those currently being used and considered the best, that is, the charts of Goos and Thornton's *English Pilot*, both clearly out of date.[49] In particular, he thanked the CIO for the use of their pilots' journals and consulted what astronomical observations there were. Again, he called for more information from navigators in order to improve further the chart and navigation, in particular, regarding the winds and currents and the location of two French possessions, the islands of Île de France and Île de Bourbon, strategic islands (i.e., islands of the Mascarenes or today the islands of Mauritius and Réunion) to the east of Madagascar. He included no sailing directions that pilots might use in consultation with the map as the map was far too general and, instead as would be his custom, he wrote of the ways he improved the map.

At the hydrographic office, what Bellin (in effect, a geographer/hydrographer of the study) presented was France's ability to produce maps and charts that served a scientific discourse of authoritative knowledge on a European stage that was populated with out-of-date maps and charts and secreted works (i.e., the Dutch) that were quickly becoming likewise obsolete. While the vast majority of Bellin's work would be at a small or medium scale, his contributions to the "geographical archive"—of which the work of the Dépôt was a part—participated in the Enlightenment ideology confident that a reasoned analysis operating on accumulated data would produce an authoritative (or *true*) work. With his works, Bellin would suggest that he challenged other chart makers to produce such works to serve the needs of the pilots and the surety of navigation. His position would be challenged by a fellow countryman and pilot of the Company.

Before turning to the works of this CIO pilot, it is important to understand the eighteenth-century distinction between British chart making and French chart making at the Dépôt. As seen earlier in this chapter, Thornton's work provided the best information available in chart form, but once his plates were purchased by the publishers, Mount and Page, it was nearly assured that the work would not be updated and was soon used as the *before* draft to be corrected. While Bellin was responsible to sell Dépôt charts to map and print sellers[50] as a means of dissemination (and, no doubt, income for the Dépôt), the plates remained at the Dépôt, an institution of the Colonial Machine. Certainly Bellin would complain about financial support, and the Dépôt (and the Navy) had their share of financial problems and lack of leadership, but as McClellan suggests the institution and bureaucratic roots were deep and wide[51] and perhaps where one institution may have faltered another succeeded. In fact, the real leap forward in the charting knowledge of the area would emerge from the cadre of officers of the CIO—another institution within the Machine.[52] The combination of skill, *in situ* chart making, developments in instrumentation, and institutional direction and support helped propel the French chart trade to a dominant position.

THE CHARTS AND MAPS OF D'APRÈS DE MANNEVILLETTE AND D'ANVILLE

The work of Jean-Baptiste-Nicolas-Denis d'Après de Mannevillette (1707–1780), the son of a CIO officer and himself a captain in the CIO, would bring greater hydrographic clarity to the region and highlight the challenges of relying on bureau map (or chart) makers like Bellin.[53] Despite Bellin's claims of using empirical data, he had to admit it was very difficult to be certain in his kind of study when one sees only with the eyes of others.[54] D'Après, an active pilot and cognizant of the practice of making observations and using instruments, had devoted himself to the collection of navigational material for the area of the East Indies since 1733[55] and perhaps believed himself to be in a far better position to assess the information available to improve existing charts and draft new ones. He was better able to present and serve the *gaze* of the active navigator with oceanic charts and pilotage charts of an appropriate scale for detail. Moreover, as a young man he had studied mathematics, geography, and astronomy in Paris with the celebrated premier royal geographer, Guillaume Delisle and astronomer, Philippe Desplaces (d. 1736) both members of the Académie to which d'Après would be awarded the position of correspondent in the 1740s. He had proven his technical abilities in testing the use of the English quadrant (or octant) on board the *Galathée* on a voyage to the East Indies.[56] With the octant, navigators would be able to make (more accurately) astronomical observations, and with the use of tables and almanacs determine their latitude. The results of that commission was a published work describing the English quadrant and its usefulness for observing latitude at sea.[57] It demonstrated his scientific abilities and sensibilities for the improvement of navigation and a connection with one of the primary institutions of the Colonial Machine, the Académie, for the coveted (and necessary) approbation for printing. As he might have expressed, he was no mere clerk or draftsmen in a Parisian office, but an active observer cognizant not only of the art of navigation but also the necessary utility of charts for navigating.

Before d'Après produced his famous *Neptune Oriental*, he was not only aware that maps needed to be improved but expressed the need for the French to free themselves from a reliance on English and Dutch works. In January 1741, d'Après wrote to Jacques Duval d'Eprémesnil (1714–1764), the director of the commerce of the Company, in order to communicate to him his observations for calculating longitude with the octant.[58] Such work would allow him to revise and correct existing works and to produce new works for the East Indies. His charts truly represented the *gaze* of the navigator not a compiler in a Parisian office or a shop along the banks of the Thames.

He continued to use his knowledge to make observations and measurements for the improvement of navigation in these Eastern waters and as a correspondent to the Académie reported his findings. D'Après drew his information from the archives in Lorient and Paris including ships' journals from Company vessels travelling to India and China as well as his own observations and experiences.[59] "My constant attention, in several voyages I have made to India and China, to all the observations that might yield any advantage, authorises [sic] me therefore to join those of my own, yet without giving the preference to my own judgment unless truth obliges me to it."[60] Having accumulated valuable information and charts by his own hand he sought permission to publish a work, *Neptune Oriental*, to which the CIO responded by requiring the Académie to examine the work. Once again with the valuable approbation from this institution as well as Louis XV and financial support for its publication from the CIO, d'Après' *Neptune*, twenty-six charts covering navigation from the Cape of Good Hope to the East Indies and China, appeared in 1745. D'Après dedicated the work to Jean-Louis Henri Orry de Fulvy (1703–1751), royal commissioner for the CIO. He praised Fulvy's protection of all those who had contributed to the good of the State, the glory of the Nation, and the surety of the commerce of the Company which a great king had committed to Fulvy's care.[61]

D'Après' preface is a brief history of the voyages and charting of the area beginning with the Portuguese whom the Dutch would replace. He acknowledged that the early charts were plain charts and did not take into consideration the curvature of the earth. The expansion of pilotage required the development of particular charts (pilotage charts) of small areas as well as larger charts covering the broader seas (oceanic charts), but also the use and acceptance of the Mercator projection. D'Après registered surprise, however, at the Dutch, whose navigation in this part of the world was so extensive and yet they confided their economic power and wealth in such imperfect works.[62] Of the charts made public of this region, Goos' general maps were considered the best. The particular maps (more detailed or larger scale maps) remained in manuscript and secret, but what had appeared to the eyes of the navigators he suggested was still surprisingly "imperfect."[63] D'Après also mentioned that Thornton's *English Pilot* was "much esteemed by navigators" because there were more particular maps for the navigator, but since 1703 little was done to update the work and the 1734 edition only added two or three new charts.[64] Consequently d'Après set out to produce his work. In making the new charts he consulted existing charts and maps and ships' journals and used three methods to improve the charts of the area: by astronomical observations; by coastal surveys using triangulation; and by noting the ship's course. After some discussion of specific charts and observations in the region, d'Après noted that some errors were produced by poor "rescaling" of charts.[65] He warned that sometimes small scale maps were enlarged, but, as they were small scale maps, there were hazards even islands that may not have appeared and now trusting a large scale map with so much information missing could be dangerous.

In addition to pointing out hazards inaccurately placed, and noting information from the Portuguese (and Dutch) rutters and the *English Pilot,* d'Après, in the accompanying *Routier* published the same year, described the winds, currents, and weather that prevailed along the coastlines such that navigation at certain times of the year was impracticable even dangerous.[66] Sailing further east, he provided advice and warnings regarding the hazards in the South China Seas particularly near the islands of Java and Sumatra.[67] Further east, however, he admitted that while he intended to provide more information regarding the navigation of the Moluccas and Philippine Islands, he was cognizant that ships were put in greater danger by "dubious" instructions and decided that it was best to not include this material until more certainty was evident. More than two decades later, Louis-Antoine de Bougainville (1729–1811), the first Frenchmen to circumnavigate the globe (1766–1769), wrote in appreciation for d'Après' "candor" and wished that all compilers of charts were as forthcoming when information was uncertain.[68] In acknowledging that French charts of this region could not be depended upon, Bougainville drew attention to the serious gaps in hydrographic knowledge.

In 1750 with the Navy's Hydrographic Office under the direction of the energetic and well-connected La Galissonière (1750–1756), d'Après was soon given the command of the Company vessel, *Glorieux*. As a part of an eighteenth-century "act of reconnaissance," that is, "the purposeful movement through and examination of the world,"[69] d'Après and his companion (for part of the voyage) were to determine the exact position of the Cape of Good Hope and the Mascarenes, to better map the southeastern coast of Africa, and to transport Nicolas Louis de La Caille (1713–1762), to the Cape to measure the arc of the meridian and observe the southern sky. An unexpected stop at Rio de Janeiro allowed d'Après to make observations and to revise the location of the port 4 degrees 7 minutes further west than its location on the maps of Goos, a discrepancy of 247 miles.[70] He also found that the Mascarenes were situated further east by 300 miles and smaller than had been reported which explained why mariners thinking they were in view of them never saw them. D'Après' observations with the octant aboard ship were confirmed by those of La Caille on land.

Upon the completion of this voyage, d'Après presented the king with a corrected chart of the eastern ocean, *Carte Réduite de L'Océan Oriental* (1753), which far surpassed what had been available from the publishing house of Mount and Page (i.e., Thornton's *English Pilot*, and the chart of Goos and its 1710 revision by Van Keulen). D'Après' chart of the eastern ocean provided better depictions and locations of the island of Madagascar, the archipelago to the north-east of this island, the Mascarenes, and points further east.[71] His chart even revealed the inaccuracies of Bellin's 1740 chart of the eastern ocean and verified that the quality of observations—supported by the Colonial Machine—made by him and La Caille were necessary for the progress of navigation. Like Bellin, d'Après certainly participated in the scientific discourse with his publications. Below the title of his chart, he included a familiar statement regarding the use of up-dated sources: "Dressée sur les Mémoires, Routiers, et Journaux des plus habiles navigateurs, Assujettis aux observations astronomiques les plus exactes à Celles de l'Auteur en particulier, et à ses Remarques. [Drawn from the memoirs, rutters, and journals of the most able navigators, adjusted by the most exact astronomical observations by those of the author in particular and on his remarks.]"[72] Here would be the distinction between the two as for d'Après, who would for much of his career be able to say that he had seen with his own eyes, he was better able to properly analyze and evaluate all of his sources. Not only had d'Après made the observations, seen with his own eyes, but when he utilized or had to rely on the measurements and observations of others, he was a much more discerning participant of the Enlightenment ideology of encyclopedism. While the works of Bellin and d'Après participated in the discourse of science d'Après might have argued that his experience provided him greater authority in that discourse.[73]

Another authoritative voice in that discourse was that of d'Anville, secretary of the duc d'Orléans and the well-respected bureau cartographer of the eighteenth century. The CIO commissioned him to produce a map of India and one of the West African coast using CIO sources. Having just produced a large map of Asia in two parts in 1751 and 1752, d'Anville reported that he was reluctant to turn his attentions to a map of India that would require a greater detail of knowledge for a region in which there remained much unknown, but having been honored by a request by the Company and recognizing the process by which maps were improved, he acquiesced.[74] For d'Anville this intellectual exercise or process was "positive geography," that is, "an extended exercise in reasoned argument" in which "[t]he geographer progressively added data into the framework of the graticule . . . [and] had to make many informed decisions about the quality and relative importance of each source in order to reconcile conflicts."[75] In this cartographic process, d'Anville was a recognized authority. In September 1752, d'Anville produced his map of India as a contribution to the improvement of geography.[76]

The map included an inset for the entrance to the Ganges River with soundings and some navigational hazards, two smaller insets of the city of Goa on the western coast of India and the city of Masulipatnam along the eastern Coromandel coast of India. The density of information along the Malabar (southwestern coast of India) and Coromandel coasts is in sharp contrast to the empty area in the interior north central region of the map where the scale of measures is printed along with a statement indicating that knowledge of this area of the country was lacking. In the upper left corner, he even directed readers interested in the area around the Indus River to examine the first sheet of his recent map of Asia. He recognized that in publishing his work, he was revealing the absences and flaws in the geographic (and hydrographic) knowledge of the region. It was a useful map for the members assembled in, say, the boardroom of the CIO, the desks or libraries of learned or curious elites; this small scale map would serve a political purpose as the CIO was constantly in need of justifying its existence (and monopoly) to the government. And yet, more than a decade later Bougainville wrote in praise of the usefulness of d'Anville's map of India (1752) during his circumnavigation. "On the whole course I have verified, by my observations, the exactness of his positions, and of the bearings which he gives to the most interesting parts of this difficult navigation. I shall add, that New Guinea and the isles of Papua come nearer the truth in this map, than in any other which I had in my hands."[77]

In 1753, d'Anville published his explanation to accompany the map, *Eclairissemens Géographiques sur la Carte de l'Inde*, which represented his familiarity and preference for ancient authors such as Ptolemy but equally his knowledge and familiarity with more contemporary cartographic/hydrographic works by Goos, Blaeu, and Thornton, Captain Cornwall's work, and d'Après' *Neptune Oriental*. While d'Anville did not always agree with d'Après' presentation of information in the *Neptune*, he did, in particular, when discussing the islands around the area of Mergui (off the coast of present day Myanmar), recommend consulting d'Après specifically for the means of navigation around this archipelago.[78] Despite the concerns over bureau map makers, the CIO and others in the field of geography respected d'Anville's thoughtful methodology. The British map and chart producers, similarly, revealed a great respect for the work of d'Anville and his methodology, as they immediately translated his map and memoir.[79]

BRITISH RESPONSE: ENGLISH TRANSLATIONS, CRITIQUES, AND PERIODICAL PIECES

With Paris the center of the map and chart trade, and the CIO having been able to compete with the EIC and approach comparable profits, harsh criticism was aimed at British maritime productions by some of its very own. Coming at the same time that John Green chastised his government for its lack of support for cartographic enterprises (see chapter 4) some action was expected.[80] At least since the publication of his 1755 pamphlet, Charles Noble had made quite clear the deficiencies and imperfections of British chart production. He compared the British and French maritime activities and when it came to charts, in a much quoted and rather entertaining bit, he had scarcely a good word to say about the state of British charts for the East Indies.[81]

> The French have been at pains, to improve their Navigation and their Charts; Those of the Indian Seas, by Monsieur D'Après de Mannevillette, a Captain in their Service, exceed every thing of the kind in Europe. It is a pity they are not translated into English, for the benefit of our Navigators. Those who understand the language, and have seen them, must have a despicable opinion, of our Indian Pilot, with which Messrs. Mount and Page have long imposed on, and picked the pockets of our Countrymen, and which are only fit for the Grocers and Chandlers Shops, or posterior uses.[82]

Understandably among the first efforts to ameliorate the situation was a translation of the best French source available, d'Après' *Neptune*; William Herbert, a purser's clerk of the EIC, translated the *Neptune* in 1758.[83] He dedicated the work to the Directors of the Company and reported that while the translation remained true to what d'Après had published, he included his comments in italics and in notes based on his examination of ships' journals, manuscript and printed charts and plans, and works he himself collected while in India. Herbert also included an appendix in which he reported accounts, directions, remarks, instructions, etc. from various pilots navigating the region but not addressed in d'Après' work. Herbert intended to continually add to this Appendix: "as new Discoveries of Dangers are sometimes made in the most frequented Parts, I propose to continue the Appendix, by inserting such Improvements from Time to Time as I may be enabled to collect."[84] Herbert then also participated in the Enlightenment ideology of encyclopedism. This translation heralded the end of Thornton's *English Pilot* as by the 1760s pilots navigating these seas simply required more up-to-date charts, charts to cover the expanded trading region, and large-scale pilotage charts.[85]

Britain also responded with its own well-connected engraver and map publisher, Thomas Jefferys. In 1754, Jefferys published his map, *The Seat of War on the Coast of Choromandel* with an accompanying memoir and dedicated the work to the Directors of the EIC. Jefferys, a royal geographer, was not a "scholar-geographer" like d'Anville, but was quite able to use information—such as that from the Board of Trade and Plantations—provided by others to defend state interests.[86] While Jefferys was well known as a defender of British colonial possessions, in particular, through his maps of America and the West Indies (see chapter 4), he appreciated and respected the well-constructed maps by d'Anville. He referred to d'Anville as "that eminent Geographer, who has made much greater Discoveries in the antient [sic] Geography of India, than any who have preceded him."[87] Jefferys did double duty in praising d'Anville's attention to geographical works and while taking a swipe at the Dutch who, according to Jefferys had at best only mercenary concerns. For example, Palliakatta was a "considerable settlement of the Dutch" with a large lake of nearly eight leagues in length, but it did not appear on their more recent maps and was only returned to the geographical image thanks to d'Anville's search of ancient sources. Jefferys wrote, "But for that Defect not the Geographers are to blame, but the indolent Dutch, who have no Inclination to improve the Science, or Curiosity beyond their Commerce."[88]

Jefferys, a commercial map publisher, intended to position British and French map and chart makers in the camp of those interested in improving science while the "indolent" Dutch, only interested in safeguarding their commerce, were content to let many a ship sail to its doom. In the travels (1768–1771) a little over a decade later for the VOC, Captain J. S. Stavorinus would come to write scathingly about the inaccuracy of charts for the seas east of Batavia.[89] "It would be well if the charts of these parts, which the Company [VOC] give to their vessels, were correct; it is this that makes the navigation so dangerous. . . . None of the islands which I met with, except the group called the *Hen and Chickens*, are laid down in their true latitudes. . . . This inaccuracy not only renders these charts useless, but likewise extremely dangerous; for instead of being, as they ought, the surest guide and dependence of the navigator, they mislead him, and become his bane." Worse yet he suggested that on account of commercial competition, "It is not impossible, but this may be purposely left so, and that it is an adopted opinion, that it is better to expose a few ships to the danger of shipwreck, than to correct errors which might operate to render the navigation towards the Spice Islands difficult and hazardous for other nations."[90] Stavorinus' comments about his voyages were only published at the end of the eighteenth century when the VOC was near the end of its existence (1795). The VOC had ruthlessly taken commercial power in the East Indies from the Portuguese and kept its particular charts for the seas east of Batavia in manuscript form and virtually under lock and key. Ships' captains were to keep a locked chest onboard in which the *secret* Dutch charts were stored and at the first signs of danger from a European competitor, the chest was to be thrown overboard lest its precious cargo fall into the hands of a trading challenger.[91]

The French were not unscathed by Jefferys' comments, as at the end of his work perhaps with some nationalist pride so evident in his works on the Americas, Jefferys did take d'Anville to task for not citing his sources. He suggested that d'Anville's position accorded him sufficient experience to know "what Errors Collectors commit; and what a jumble they often make of Things for want of either Care or skill to examine discording Relations." Jefferys continued in a more damning tone.

> This Method, commonly practiced to conceal an Author's Laziness or Ignorance, is often made-use of to convey Falsehood more securely, by rendering the Detection difficult. For these Reasons, this Mal-practice ought to be exploded; and every such Writer considered as a Suppressor of Evidence, which is near a-kin to Forgery. In short, a History without vouchers ought to pass for a History of the Compiler's own Invention, and not a collection from others. No Writers so guilty of this Enormity as the *French*; and Mr. *D'Anville*, we fear, has suffered from the same Misbehavior of his Countrymen more than once.[92]

While few map or chart makers at the time could stand on superior ground with respect to the use of sources, Jefferys' commentary reminded of the mid-eighteenth-century Enlightenment ideal of the geographical practice—"an extended exercise in reasoned argument."[93] Printed works accompanying the maps or charts legitimated the process and were to convince readers of the authority of the maps or charts they viewed. Beginning in the 1740s a British readership of elites and the middling sort were similarly informed about the events surrounding the production of these charts and their importance.

In the same year a broader audience of the middling sort was able to read about the region and the military contests and view lovely prospects and informative maps. The *London Magazine* (1754) included several descriptions and prospects of key locations such as Bombay, Batavia, and Fort St. George (Madras) and even a map, *A Map of Coromandel Shewing the Seat of War on that Coast*, printed for Robert Baldwin depicting the battle sites, and sieges along the southeastern coast of India, an economically valuable marketplace and politically challenging region.[94] In the November 1760 edition of the *Gentleman's Magazine* the editors printed a report of the King's Speech to the Houses of Parliament. With regard to the war effort in the East Indies, George III, who had only just taken the throne upon the death of his grandfather reported: "Our advantages gained in the East Indies have been signal; and must greatly diminish the strength and trade of France in those parts, as well as procure the most solid benefits to the commerce and wealth of my subjects."[95] By the close of 1760 Britain had the upper hand in the East Indies and the EIC would soon become not only a powerful trading company in the region, but a colonial administrative agency, an arm of the imperial machinery.[96] In the June 1760 edition of the *London Magazine*, the editors reminded their readers that they had provided them with a map of the Coromandel Coast in the 1754 volume (after page 440) and now included "a map elegantly engraved, of the rest of the Northern Coast of that Province." Given the events of the past few years, they wrote, "[m]aps which must now be very necessary for our readers [sic] inspection. . . . Our readers may consult, besides our last Volume and the present Year, our General Index . . . where they will find every thing relating to this coast."[97] The following year (in July) the editors listed the EIC's forts and factories in the East Indies and the South coast of China and accompanied the list with a map by Thomas Kitchin, *A General Map of the East Indies and That Part of China Where the Europeans Have Any Settlements or Commonly Any Trade*.[98] Two years later with the close of the Seven Years' War in the June 1763 edition an account on the Philippine Islands and the first chart of these Islands appeared with a map of Manila, a site taken from the Spanish by the British the year before.[99] An informed readership supplied with such sources not only witnessed a greater British intrusion into this area as it searched for new markets, but also grew confident in its ability to know distant lands. This would be increasingly important as British interests post-War of American Independence would begin to shift from the Atlantic to the Indian Ocean.[100]

In the final year of the war, a French ship of the Company, the *Dromodaire*, wrecked (February 1762) off the coast of the Cape Verdean island, St. Vincent. D'Après, who by this time had left the seas, took up the post of inspector at the newly established Hydrographic Office of the Company in Lorient.[101] The naval minister asked d'Après to complete sailing directions for those vessels that would voyage to the East Indies so that such tragedies could be avoided. D'Après presented his *Mémoire* to the king and attached to the work a chart of the Cape Verde Islands, *Carte des Isles du Cap-Verd dressée sur les remarques et les observations les plus exactes des navigateurs*, on which he marked the "dangerous canal" between the Cape Verdean islands of St. Vincent and St. Lucie.[102] Upon presenting this work, d'Après expressed to the king his renewed desire to begin a new edition of his *Neptune* and received great encouragement for his work. As a sign of his appreciation, the king awarded d'Après the Order of St. Michael and an annual pension of 1200 livres. D'Après reported years later in the preface of his *Instructions* (1775) that the king's appreciation and generosity motivated him to continue to work towards the perfection of the navigation of the East Indies.[103]

D'Après and Bellin would continue to produce works for the region even though by 1762 d'Après was, in effect, now a bureau chartmaker like Bellin though d'Après would never suggest *like* Bellin.[104] Bellin, being in Paris or Versailles, was not well placed to communicate with CIO officers and at best could only participate in the bureaucratic flurry of papers with the office of the Marine. Even after the suspension of the CIO in 1770, d'Après retained his position and pension until his death in 1780 at which time the CIO archives were moved to the Paris Dépôt.[105] During his tenure at Lorient, d'Après would argue that his years of experience (more than forty) at sea allowed him a greater understanding of the sources that he was using to produce updated charts as well as to ask the pertinent questions. He was better placed being in Lorient to communicate with CIO pilots and had become a well-respected authority for the charting of the eastern seas. Among his colleagues in the maritime brotherhood, he could count on the information from the talented British hydrographer, Alexander Dalrymple (1737–1808).

THE WORK OF ALEXANDER DALRYMPLE

Dalrymple had entered into the service of the East India Company at a young age and by 1757 found himself in the Secretary's office of Fort St. George (Madras) just at a time when the Seven Years' War was spilling over into the East Indies and the Company was intent to find not only new markets to solidify their control in the broader region, but also a more strategic port of operations.[106] For at least five years, Dalrymple had observed the political, economic, military, and navigational issues in the region, and he recognized that other trading stations and opportunities needed to be found. The French had laid siege (again) to Madras at the end of 1758 and early 1759 making new ventures all the more necessary. Madras had no harbor so ships had to lie offshore one or two miles, and when the north-east monsoons arrived the Coromandel coast was not safe for shipping, and along the northeastern coast, the harbor at Bombay was not ideal either.[107]

Fired up by the success of William Wilson commanding the *Pitt* who opened a route to China (1758–1759) available at any time of the year in contrast to the trade routes determined by the monsoon season, Dalrymple proposed to seek new markets in the islands of the Sulu Sea on board the *Cuddalore*.[108] He arrived in Macao in early July 1759, and, while waiting for fresh ships from Europe, he investigated the islands north of the Philippines and part of the coastline of Cochin China to Tourane. This was his first opportunity to gather first-hand information and observe native and British relations in the field, and the subsequent frustrations, dangers, and diplomatic intrigues ever present in this region. Returning to Canton, Dalrymple found himself arriving at an auspicious moment (depending on one's point of view) as five East Indiamen were ready to leave China, but the French having recently taken control of the southwest coast of Sumatra posted vessels in the Sunda Strait and were lying in wait for such prized possessions.[109]

At the end of 1760, using copies of Dutch maps of the Celebes and information gathered from the voyage of the *Pitt*, Dalrymple threaded his way through the islands and stopped at Sulu long enough for him to make a preliminary trade agreement with the Sultan of Sulu and then on to the island of Sumbawa completing his task of bringing the East Indiamen through the uncharted waters. He continued his voyages throughout these islands of Indonesia conducting surveys and gathering information that few vessels had before. Finding the Dutch hostile on Macassar, he was pleased on Mindanao to meet a kindred spirit, Don Manuel Galves, the Spanish Governor, who shared with him knowledge of the local waters.[110] From May 1759 to early 1762 he travelled the Southeast Asian seas making surveys, taking soundings, gathering what knowledge of the waters he could and making contact with the Sultan of Sulu to establish trade relations. At nearly twenty-five years old, Dalrymple had completed his first trading venture and had had the good fortune to find himself in uncharted or little known waters and, more importantly, the talent to gather essential information for the improvement of navigational charts. "The voyage had provided Alexander Dalrymple with that grounding in navigation, seamanship and surveying which was to enable him to become the first great hydrographer in British history."[111] Despite the work of several pilots (mostly from the EIC), England's reputation for chart making still suffered as navigators were still served by English translations of French works.

As noted in chapter 3, in 1770 James Cook, hydrographer and naval captain, castigated both pilots and publishers for producing untrustworthy charts: "between the one and the other we can hardly tell when we are possessed of a good Sea Chart until we our selves have proved it."[112] He accused navigators of making up soundings and charting coasts they had never seen, while publishers printed these so-called surveys as accurate with no attribution to sources/authorities for verifiability.[113] In one of his many early works Dalrymple amplified Cook's critique and initiated a common concern and plea, that is, pilots needed to know how to make observations with the best of intentions and to share their observations with such chart-makers as himself in order to improve the safety of navigation. In his *Essay on the Most Commodious Method of Marine Surveying* (1771), he explained that his intention in publishing such a work was "to assist and exhort the Former [pilots]; to encourage the Diffident; to shame the *Careless*, by showing *with how little Trouble* useful observations may be made; and to instruct the Ignorant in the Practice of a few useful Rules."[114]

In 1771 and 1772 Dalrymple published several works that provided the groundwork for the dominance of the British in hydrography (if not geographical production in general). His memoirs on the charts of the China Seas and the West Coast of Pulawan as well as his *General Introduction to the Charts & Memoirs* (1772) made public what information and charts there were for areas little known or for which more detailed information was necessary. He wrote, "I flatter myself the Navigator will reap advantage from the information these Charts convey; and I may venture to say, nothing yet published of those parts are equal to them."[115] He did caution the reader/viewer to remain vigilant as over confidence in navigating these waters could lead to disaster. In particular for the water around Pulawan, "Ignorance is always confident; there are many seamen who never made a Draught in their lives, who pretend to be able to guess Distances exactly but how they can determine their Guess I have no Conception, as the Irregularity of the Tides and Currents near Land make the log a very uncertain Guide."[116]

In the decade to follow, Dalrymple compiled all the information available from the EIC and, in effect, acted as the Company's hydrographer without the position or title. This work led to his *Charts and Memoirs* (1772). In correspondence with d'Après since the mid-1760s, Dalrymple sent a set of his new charts to d'Après in time to assist the French hydrographer in revising and adding charts to the re-edition of his *Neptune Oriental* (1775) which not only benefited from Dalrymple's contributions, but also from information on the shortest route from the Mascarenes to India from the travels of Jacques-Raymond Grenier (1767/70), observations from Bougainville's circumnavigation (1766/69), as well as the ships' journals from other pilots. Dalrymple had great respect for d'Après' work and wrote that unlike many others of the age he was truly deserving of the title, hydrographer, and even inscribed a map to him, *A Chart of the China Sea inscribed to Monsr. D'Aprés [sic] de Mannevillette the ingenious author of the Neptune Oriental: As a tribute due to his labours for the benefit of Navigation and in acknowledgement of his many signal favours to Dalrymple.*[117] The chart included a note for the abbreviation of a dozen British and French vessels whose routes were included on the chart; as well as information from a Portuguese manuscript chart, and sources for the position of two shoals. The chart aimed to be purely informative but kept elements of the printed explanation that often accompanied eighteenth-century maps *on* the chart so as to indicate its authoritative origins.

In 1773, the EIC gave 100 guineas to a new project, *Collection of Plans and Ports in East India*, for which Dalrymple intended to provide navigators with a pilot book of plans but no general coasting charts.

> It is often of the utmost benefit, and always very satisfactory, to Navigators to be possessed of *particular plans* of the Places whereto they may, by chance or accident, be carried, as well in as out of the common track; I have therefore thought it would be an acceptable service to the Publick to *engrave* the plans of *Bays, Harbours, &c. in the East-Indies*, which are in my Collection. None of these Plans being from my own observations, I cannot vouch for their exactness; but although they ought not to be *blind-fold* and implicitly relied on, they may be of very great use and assistance.[118]

As a contribution to the "geographical archive" Dalrymple acknowledged the drawbacks at this still early stage of gathering information. He presented these plans knowing that there may have been flaws as they were compiled from a variety of sources rather than as part of a systematic hydrographical survey. In addition to the one hundred copies for the EIC, Dalrymple hoped for one hundred more subscribers but only ten individuals had registered.[119] The work published in six parts over the course of a year (February 1774 to March 1775) was a series of plans of ports (eighty-three plates) from South Africa into the China Seas with sailing directions and topographical descriptions which yielded an impressive tome of more than 250 pages.[120] With an objective to improve hydrographic knowledge similar to d'Anville's actions to print his map of India (1752), Dalrymple published these plates in the hopes that more information would arrive from EIC pilots (and others) to improve subsequent works.

Dalrymple may have been disappointed at the number of subscribers but was hopeful that the support of the EIC would encourage more (and better) information to make its way from the East Indies to the archives of the India House. "I am happy to see that several draughts have lately been sent home from thence [Bombay], and I flatter myself the encouragement given by the Court of the Directors towards these publications will be an inducement to The Company's Servants abroad to transmit in future other Charts and Remarks when they see their labours may be of service to the Community."[121] He expressed further indebtedness to d'Après for ships' journals of the *Camel* and *Elephant* who had in 1760 "touched at Yulinkan on the S. Coast of Hainan."[122] He also explained one of the more unorthodox ways that hydrographic information was transferred to him. He used a chart by Van Keulen to draw part of Bongo Bay on the island of Mindanao, but the information for the other part of the island had to be retrieved from the sea where the Spaniards tried to dispose of it. "Part of Bongo Bay is taken from Van Keulen, but part of it is from a Spanish Ms. Sent by an English Renegado Norton Nicholls, or, as he then was pleased to stile himself, Nicolas Norton, from Manila on board the *Trinidada* for Spain. The Spaniards threw it overboard, but it was taken out of the sea by some of the people belonging to his majesty's ship Argo, Capt. King, who was so obliging to give me a copy of it in India."[123]

In 1779 with the loss of the East Indiaman, the *Colebrooke*, Dalrymple renewed his calls for a position of hydrographer with the same description he proposed back in 1766, that is, to examine collected material of the EIC and produce charts for the aid and security of navigation.[124] In his Memorial for the position he wrote of the desperate situation for better and more charts and that despite the work of d'Après much more was needed. This time his proposal met with success. After one year on a trial basis in which Dalrymple examined nearly 1200 ships' journals and studied the routes of various ships, the hazards noted by various pilots as well as the currents and winds in the region, Dalrymple became the hydrographer of the EIC at the India House in Leadenhall Street, London. In 1779 when Dalrymple was appointed hydrographer to the company with the requirement that fifty prints go to the Company (raised to one hundred in 1780), he had already established the process with his *Collection* project of 1773.[125] "The provision for chart publication was that Dalrymple's expenses in engraving and printing 100 copies of any chart he chose to compile would be met by the Company, after which the plate became his property and further impressions to use or sell as he pleased were at his own expense of paper and presswork. The editorial control was entirely in his hands, as was planning for publication. Market forces were conspicuously absent."[126] In other words, Dalrymple did not need to concern himself with the commercial viability of any chart that he chose to compile and print as, say, the publishers, Robert Sayer and John Bennett, would.

Almost immediately the knowledge he had gathered was of immediate use for the economic and military pressures caused by the War of American Independence. In 1780, Britain declared war on the Dutch which left the trade routes to China vulnerable to Dutch and French forces. Dalrymple was asked to investigate the so-called Pitt's Passage and to provide charts for this alternative route which yielded in 1782, *Memoir Concerning the Passages to and from China*. When the work was published it appeared only in a few prints and was issued by the Secret Committee of the EIC and specifically to the pilots of the China seas who sought safer routes than the popular passages through the Straits of Malacca and Sunda.[127] He added to this work a report on passages "late in the season" to India and China and after the War both were published in 1785 and 1788.[128] By 1783 with the return of peace, Dalrymple re-energized the efforts to seek more information for the areas further east, to find alternatives for provisioning along the route(s), and chastised pilots anew (a decade after his first request to them) to make charts.[129] "'They are all ready,' he continued, 'to exclaim against the badness of the charts made by others, indeed with sufficient reason, since those charts are constructed from the execrable materials afforded by their Observations. Were the practice of Hydrography Universal amongst Navigators, the Improvement of Charts would rapidly approach to perfection.'"[130] For the surety of navigation, the science of hydrography simply could not continue to rely on a few gifted and committed pilots. While Dalrymple recognized the infrastructure necessary "for a system of standard charts for navigation" and did receive information from Company pilots and local pilots, their work was limited in scope and fell far short of the larger hydrographic survey required.[131] While he continued to implore that all navigators work towards the "perfection" of navigation, a broader institutional structure was necessary, perhaps an Imperial Machine. Moreover, with the end of hostilities of the War, the EIC was able to continue its new sovereign role in the Indian subcontinent, and its growing political power in the area meant not only a greater number of East Indiamen on the seas and the need for new routes to expand the British trading dominance, but greater carrying capacity and therefore higher monetary stakes.

LATE EIGHTEENTH-CENTURY FRENCH ACTIVITY IN THE EAST INDIES

About the same time the French East India Company, which had been suspended in 1770, was reconstituted with a monopoly for Asian trade in 1785 and, to the chagrin of the British, was enjoying an alliance with the Dutch. The naval ministry also launched several exploratory missions with scientific (and economic) motivations into the Indian and China Seas. The future director of the Dépôt (1795), François-Etienne de Rosily-Mesros (1748–1832), was sent to examine the eastern coast of the Indian Ocean from Madagascar to the Persian Gulf.[132] Jean-François de Galaup de La Pérouse (1741–c. 1788) began what would be a fateful expedition to the coast of China and points south never to be heard from again. M. de Richery in 1785 "was instructed to chart the coasts of Tonkin, Cochin China, Siam, and Malacca" and would be sent to Cochin China again in 1786 to locate the young prince, Nguyen Anh (see below) and "to investigate the resources, trade and harbours of Cochin China."[133] Joseph-Antoine d'Entrecasteaux (1739–1793), commander of the French fleet in the East Indies, sent to explore the China Seas (and later to search for the La Pérouse expedition) described the British dominance in the region. "The more one sails in these seas, Monseigneur, the more one is astonished by the enormous power of the English in these vast and fertile countries."[134] Still France intended to compete. Taking command of the vessel, *La Resolution*, d'Entrecasteaux opened up a route to Canton through the Sunda Strait and the Moluccas during the monsoon season. In a letter of 8 February 1787, he commented on the state of knowledge on charts of this region. He wrote that he was sending a map of the route taken by *La Resolution* through the Straits of Macassar, Gilolo, and Pitt. He warned that the Dutch maps of this region—recall Stavorinus' comment on Dutch maps—were so faulty as not to be trusted and those who followed after him and used his charts would have an easier time.[135]

Among the last French hydrographic projects of the eighteenth century in the eastern seas was that instigated by Pierre-Joseph-Georges Pigneau de Béhaine (1741–1799) (Bishop of the Seminary of Foreign Missions), who in protecting the young prince, Nguyen Anh (future emperor Gia Long of Vietnam), from a Tay Son rebellion, negotiated a treaty in November 1787 with Louis XVI at Versailles. In return for French military assistance, Nguyen Anh would accord French traders and vessels favorable commercial status which would, in turn, Pigneau would argue, restrain the economic power of the British in the region. This treaty simply needed to succeed. Nearly before the ink was dry on this Treaty of Versailles and on account of the desperate domestic economic situation in France, Louis XVI (without informing the bishop) backed away from supporting the treaty and left the ball in the court of Thomas Conway, comte de Conway (1735–c. 1800), the governor of Pondicherry, to decide the level of assistance or even if the treaty should be carried out. When he found out, Pigneau was furious over the duplicity of the court and set about raising funds from various French merchants of the area. He was intent that Nguyen Anh succeed in defeating the Tay Sons and honor the commercial agreements of the treaty even if the crown was unwilling to do so.[136]

Conway sent Guy-Pierre de Kersaint (1747–1822) to bring arms to Nguyen, meeting a portion of the aid promised but not to Pigneau's liking. While France seemed (at least at this time) unwilling to embroil itself in political affairs, it still had concerns over the detail of hydrographic knowledge in the area. Kersaint was, therefore, asked by Conway to conduct some hydrographic work (August 1788) for the production of maps of the coastline and area of Cochin China (present day Vietnam).[137] Similarly, Rosily-Mesros was in July and August of 1789 conducting further hydrographic work in the region.[138] These efforts along with the valuable charts drawn by Charles-François Beautemps-Beaupré, who accompanied d'Entrecasteaux on the expedition (1791) in search of La Pérouse would bring greater authority to the charts of this region. The upheaval of the French Revolution and the execution of Louis XVI led to the end of that expedition and with d'Entrecasteaux dead and the vessel taken by the Dutch as repayment on a debt, the fate of these valuable charts was uncertain.[139] In 1795, Rossel, the second in command and in possession of much of the material from the expedition including the charts, was making his way back to France on board a Dutch ship soon captured by the British. These captured charts were soon in possession of Dalrymple at the recently established Hydrographic Office of the Admiralty, but he upheld the principal that these charts produced under the orders of a scientific expedition were protected in time of war. He refused to publish the works until they were made public in France. Ironically they were published in 1807 but news had not reached London, and coupled with Dalrymple's stormy relations with the Chart Committee of the Admiralty, he was soon dismissed from his post.[140]

CONCLUSION

In Britain, while the private trade continued to enjoy success producing maritime charts, in the last decade of the eighteenth century with the ability to establish one's longitude at sea (by use of the chronometer or through the lunar distance method with the *Nautical Almanac*) and, therefore, to conduct more successful marine surveys the faultiness of commercial charts were more and more evident as was the inability of the commercial trade to meet the demands and standards the field would require. In 1795 the Admiralty set up its Hydrographic Office and named Dalrymple, the EIC's hydrographer, to the position, to "organize" the material already gathered and publish it for the royal ships.[141] The commercial trade was still dominant and the bluebacks of these private companies were still quite popular, but the field finally began to have an official directing institutional body—part of what would soon become an Imperial Machine—one of whose goals was the surety of navigation.[142] A more official approbation came in the first decade of the nineteenth century, as a committee of the Admiralty drew up (from a total of one thousand charts) a list of two hundred "approved" works (see chapter 3).[143] More importantly for Britain, as its move to a colonial power in India became more certain, an Imperial Machine would seek more authoritative maps and charts for its "jewel in the crown."

In France similar concerns for the quality of cartographic (and hydrographic) productions prompted two royal decrees in 1773 and 1786 (see chapter 3): the first gave exclusive control to the Dépôt for the production of hydrographic charts and the second required that the chancellor approve the publication of all geographical works. A similar decree was issued in 1814 to control the dissemination of hydrographical material and coincided with the creation of a hydrographic engineering corps.[144]

While the Dépôt fell under the able leadership of Beautemps-Beaupré, the Colonial Machine was severely curtailed, and with the decline of French influence in India in the early nineteenth century there was no new charting of the region by the French. France's increased attention and a redirection of trading interests and inevitable political intervention in modern day Thailand and Vietnam, would lead to more mapping and charting in the second half of the nineteenth century. By mid-century, however, British Admiralty charts had become the gold standard for navigation and other nations now looked to emulate them.

The success of scientific expeditions, the increased commercial competition, political involvement, and military activity stimulated the demand for reliable charts. For decades there had been calls from both sides for a greater attention to charting by navigators and the continued support or the development of directing institutions. The personal state had long ago developed into the *enlightened* eighteenth-century bureaucratic state which was at least by the late eighteenth century in Britain and France coming to be well-served by geographic productions that defined territory for administrative offices at home and abroad. Hydrographic productions were required to guide their vessels from those home offices to colonies and trading interests abroad and by the late eighteenth century institutional developments had set both countries on the path to the modern age of hydrography.

NOTES

1. United East India Company or Verenigde Oost-Indische Compagnie, VOC was established in 1602. The English East India Company was established in 1600. See Om Prakash, "The English East India Company and India," in *The Worlds of the East India Company*, ed. H.V. Bowen, Margarette Lincoln and Nigel Rigby (Rochester, NY: D. S. Brewer, 2002), 1–17; and P. J. Marshall, "The English in Asia to 1700," in *The Oxford History of the British Empire: The Origins of Empire,* ed. Nicholas Canny (Oxford: Oxford University Press, 1998), 264–85. The *Compagnie des Indes Orientale* was established by Louis XIV in 1667. See Philippe Haudrère, *Les Compagnies des Indes Orientales* (Paris: Éditions Desjonquères, 2006); and Philippe Haudrère, *La Compagnie française des Indes au XVIIIe siècle*, tome I, 2nd edition (Paris: Les Indes Savantes, 2005).

2. Matthew H. Edney, "Reconsidering Enlightenment Geography and Map Making: Reconnaissance, Mapping, Archive," in *Geography and Enlightenment*, ed. David N. Livingstone and Charles W. J. Withers (Chicago: University of Chicago Press, 1999), 186–93.

3. See James E. McClellan III and François Regourd, *The Colonial Machine: French Science and Overseas Expansion in the Old Regime* (Turnhout, Belgium: Brepols Publishers, 2011), 485. McClellan and Regourd cite from the work of John Gascoigne in which he suggests that a more directing impulse was created (in 1778) with Joseph Banks' presidency of the Royal Society. See John Gascoigne, *Science in the Service of Empire: Joseph Banks, the British State and the Uses of Science in the Age of Revolution* (Cambridge: Cambridge University Press, 1998).

4. See Prakash, "The English East India Company," 2–3. The EIC operated with a charter from the government for a monopoly of trade in the east from 1600 to 1833 at which point the Company became a colonial administrative body in India. For a brief history see, Kenneth R. Andrews, *Trade, Plunder and Settlement* (Cambridge: Cambridge University Press, 1984), 256–79. Also see Marshall, "English in Asia."

5. Andrews, *Trade, Plunder*, 262.

6. Andrews, *Trade, Plunder*, 264.

7. Marshall, "English in Asia," 275.

8. Marshall, "English in Asia," 282.

9. Marshall, "English in Asia," 283.

10. Andrew S. Cook, "Establishing the Sea Routes to India and China: Stages in the Development of Hydrographical Knowledge," in *The Worlds of the East India Company*, ed. H.V. Bowen, Margarette Lincoln and Nigel Rigby (Rochester, NY: D. S. Brewer, 2002), 121.

11. See H. V. Bowen, "'No Longer Mere Traders': Continuities and Change in the Metropolitan Development of the East India Company, 1600–1834," in *The Worlds of the East India*

Company, ed., H. V. Bowen, Margarette Lincoln and Nigel Rigby (Rochester, NY: D. S. Brewer, 2002), 19–32.

12. Cook, "Sea Routes," 121. "The need to repeat voyage patterns year after year, and to make predictable visits to a series of ports, necessitated a formal knowledge of currents, coastlines and directions for sailing to them." In England however there was "no formal mechanism for official chart publication, and [the EIC] generally exercised looser control over shipping and routes."

13. Haudrère, *Les Compagnies des Indes Orientales*, 69–71.

14. McClellan and Regourd, *Colonial Machine*, 54. "The crown controlled the Compagnie through the shares it held, through its ability to appoint directors, and through a royal commissioner serving de jure on the governing board. There is no question that the Compagnie des Indes was a state-chartered institution closely monitored and controlled by the crown."

15. McClellan and Regourd, *Colonial Machine*, 54.

16. François Charpentier, *Discours d'un fidele sujet du roy, touchant l'establissement d'une compagnie françoise pour le commerce des Indes Orientales: adressé à tous les François* (Paris, 1665)

17. Charpentier, *Indes Orientales*, 6.

18. Marshall, "English in Asia," 283–84; and McClellan and Regourd, *Colonial Machine*, 56.

19. Since the late seventeenth century the *Archives de la Marine* was housed in the monastery of the Petits Pères near Place des Victoires, Paris. In 1681, Colbert had instructed pilots to deposit their ships' journals to the clerk of the naval office. These portfolios of documents, plans, and charts would become the foundation (in 1720) for the Hydrographic Office or Dépôt with Jacques Nicolas Bellin as the clerk of the Dépôt and hydrographer. See Olivier Chapuis, *A la mer comme au ciel* (Paris: Presses de l'Université de Paris-Sorbonne, 1999), 159. The CIO would not have its own archive until the 1760s in Lorient. In London since at least the second quarter of the eighteenth century, EIC ships' captains were ordered to deposit their journals at the East India House upon arriving from their travels. See Cook, "Sea Routes,"122.

20. Cook, "Sea Routes," 122. Also see T. R. Smith, "Manuscript and Printed Sea Charts in Seventeenth-Century London: The Case of the Thames School," in *The Compleat Plattmaker*, ed. Norman J. W. Thrower (Berkeley: University of California Press, 1978), 50–51.

21. Sarah Tyacke, "Geography is Better than Divinity: The Practitioners' Story" (Sandars Lectures: Conversations with Maps, Cambridge University Library, 6 March 2007), http://www.lib.cam.ac.uk/sandars/tyacke2.pdf, 19, (accessed 7 June 2014); and Alistair Simon Maeer, "The Cartography of Commerce: The Thames School of Nautical Cartography and England's Seventeenth Century Overseas Expansion" (PhD diss., University of Texas at Arlington, 2006), 110.

22. Chapuis, *A la mer comme au ciel*, 159.

23. John Thornton and Joel Gascoyne, *A Plat of the Indian Sea* (London, c. 1684). See BN, SH Div. 3, Pf 213, 16. An edition (c. 1730) published by Mount and Page is found in the Map Collection of the British Library, BL, K. Mar. VI. 9.

24. Marshall, "English in Asia," 280.

25. John Thornton, *A Mapp of the Greate River Ganges as it emptieth it selfe into the Bay of Bengala taken from draught made uppon [sic] the place by the agents for the English East India Company never before made publique* (London, 1685), BN, Cartes, GE DD 2987 (7036).

26. John Thornton, *A New Map of the Island of Bombay and Sallset* (London, 1685), BN, Cartes, GE DD 2987 (6879).

27. Smith, "Thames School," 83.

28. Monique de la Roncière, "Manuscript Charts by John Thornton, Hydrographer of the East India Company (1669–1701)," *Imago Mundi* 19 (1965): 47; Also see Rodney W. Shirley, "The Maritime Maps and Atlases of Seller, Thornton, and Mount & Page," *The Map Collector* 73 (1995): 2–9.

29. See Roncière, "Manuscript Charts" and Shirley, "Maritime Maps."

30. In his doctoral dissertation, "Cartography of Commerce," Maeer suggests that Thornton's manuscript maps included flags at those locations that belonged to the particular trading company as an indication that commerce was the purpose (116).

31. John Thornton, *The English Pilot, the Third Book on Oriental Navigation* (London, 1703), BL, Maps C. 22. d. 30.
32. Thornton, *English Pilot*, dedicatory page.
33. Henry Cornwall, *Observations upon Several Voyages to India Out and Home* (London, 1720).
34. Susanna Fisher, *The Makers of the Blueback Charts: A Brief History of Imray Laurie Norie & Wilson Ltd.* (Ithaca: Regatta Press Limited, 2001), 3–4.
35. Philippe Haudrère, "The 'Compagnie des Indes' and Maritime Matters (c. 1725–1770)," in *Ships, Sailors and Spices: East India Companies and Their Shipping in the 16th, 17th, and 18th Centuries*, ed. Jaap R. Bruijn and Femme S. Gaastra (Amsterdam: NEHA, 1993), 81.
36. F. Dassié, *Le Routier des Indes Orientales et Occidentales* (Paris: Jean de la Caille, 1677).
37. "An Account of Four Books: II, L'Architecture Navale, avec le Routier des Indes Orientales & Occidentales par le Sieur Dassié, " *Philosophical Transactions*, no. 135, May 26, 1677, 879–80.
38. "An Account," 880; and Dassié, 14.
39. Dassié, 70.
40. P. J. Marshall, "The British in Asia: Trade to Dominion, 1700–1765," in The Oxford History of the British Empire: The Eighteenth Century, ed. P. J. Marshall (Oxford: Oxford University Press, 1998), 490–91.
41. Cook, "Sea Routes," 126; see also Haudrère, "'Compagnie des Indes' and Maritime Matters," 81; and Marshall, "British in Asia," 488.
42. Cook, "Sea Routes," 126; and Haudrère, "'Compagnie des Indes' and Maritime Matters," 82.
43. Cook, "Sea Routes," 126; and Haudrère, "'Compagnie des Indes' and Maritime Matters," 82.
44. Andrew S. Cook, "Surveying the Seas: Establishing the Sea Routes to the East Indies," in *Cartographies of Travel and Navigation*, ed. James R. Akerman (Chicago: University of Chicago Press, 2006), 71–72.
45. Cook, "Surveying the Seas," 71–72.
46. McClellan and Regourd, *Colonial Machine*, 60.
47. McClellan and Regourd note that there was a close connection between the merchant marine and the Royal Navy (54).
48. Jacques Nicolas Bellin, *Carte Reduite de l'Ocean-Oriental ou Mers des Indes* (Paris, 1740), BN, Cartes, GE DD 2987 (9661B).
49. Jacques Nicolas Bellin, *Observations sur la construction de la carte de l'Ocean Orientale ou Mers de Indes* (Paris, 1740), AN, Marine, 3JJ 4, Section 13.
50. Mary Sponberg Pedley, *The Commerce of Cartography: Making and Marketing Maps in Eighteenth-Century France and England* (Chicago: University of Chicago Press, 2005), 79.
51. McClellan and Regourd, *Colonial Machine,* 49–167. In Part One, McClellan and Regourd describe the various layers of bureaucratic institutions involved in the Machine.
52. McClellan and Regourd, *Colonial Machine*, 142–47.
53. For information on the life and career of d'Après de Mannevillette, see Manonmani Filliozat, "D'Après de Mannevillette, Captain and Hydrographer to the French East India Company (1707–1780)," *Indian Journal of History of Science* 29, n. 2 (1994): 329–42; Manonmani Filliozat, "J. B. D'Après de Mannevillette et la Cartographie de la Nouvelle Route des Indes," *CFC*, 175 (2003): 6-16; Manonmani Filliozat, "*Le Neptune Oriental*: Une Somme de la Cartographie de la Compagnie des Indes," *Cahiers de la Compagnie des Indes* 3 (1998): 21–30; and a collection of chapters edited by Eric Saunier, ed., *Autour de D'Après de Mannevillette: Savant Navigateur Havrais au siècle des Lumières* (Havre: Imprimerie la Petite Presse le Havre, 2008).
54. Jacques Nicolas Bellin, *Description Géographiques des Isles Antilles possédées par les Anglois* (Paris: Imprimerie de Didot), 1758), 42.
55. Chapuis, *A la mer comme au ciel*, 249.

56. See Claude Briot, "L'Apport de d'Après de Mannevillette dans les progrès de la Navigation en Mers des Indes au XVIIIe siècle," in *Autour de D'Après de Mannevillette: Savant Navigateur Havrais au siècle des Lumières*, ed., Eric Saunier (Havre: Imprimerie la Petite Presse le Havre, 2008), 40–42.
57. D'Après de Mannevillette, *Le Nouveau Quartier Anglais, ou description et usage d'un nouvel instrument pour observer la latitude sur mer* (Paris: Imprimerie Royale, 1739).
58. Briot, "D'Après de Mannevillette," 42.
59. Briot, "D'Après de Mannevillette," 42-46.
60. Jean Baptiste d'Après de Mannevillette, *Le Neptune Oriental, our Routier General des cotes des Indes Orientales et de la Chine. Enrichi de cartes hydrographiques tant generals que particulieres, pour server di'instruction a la navigation de ces differentes mers* (Paris: Imprimerie Royale, 1745), xxx.
61. D'Après de Mannevillette, *Neptune*, dedicatory page. "La protection que vous accordez à tout ce qui peut contribuer au bein de l'Etat, à la gloire de la Nation, & à la sûreté du commerce d'une compagnie qu'un grand Roi a commis à vos soins, me fait prendre la liberté de publier cet Ouvrage sous vos auspices; l'étendue de vos lumieres vous en a déja fait connoître toute l'utilité." He continued in the dedication to say that the French had been relying on English and Dutch works. It was necessary to use these works at the beginning being that these seas were frequented by their vessels, but our own experience has led us to see a great deal of errors on their charts. For the utility of navigators he had produced these charts. He has taken twelve years a great deal of care, and Fulvy was supportive of his first work on the English Quadrant. Your support and the support of the Commissioners led me to continue my work with great ardor and to correct our charts. Hopefully this work will be of great service to the Company. The sailing directions were also published separately the same year as *Le Routier des Côtes des Indes Orientales* with the approbation and privilege of the king.
62. D'Après de Mannevillette, *Neptune*, xxxi.
63. D'Après de Mannevillette, *Neptune*, xxxi–xxxii.
64. D'Après de Mannevillette, *Neptune*, xxxii.
65. D'Après de Mannevillette, *Neptune*, lvii–lviii.
66. D'Après de Mannevillette, *Routier*, 59.
67. D'Après de Mannevillette, *Routier*, 114–29.
68. Louis-Antoine de Bougainville, *Voyage Round the World*, trans. John Reinhold Forster (London: J. Nourse, 1772), 186–87.
69. Edney, "Reconsidering Enlightenment Geography," 175–78.
70. Briot, "D'Après de Mannevillette," 47–51.
71. Haudrère, "'Compagnie des Indes' and Maritime Matters," 86–87.
72. Jean Baptiste d'Après de Mannevillette, *Carte réduite de L'Océan Oriental* (Paris, 1753). See BN, Cartes, GE C 4644.
73. McClellan and Regourd, *Colonial Machine*, 190.
74. Jean Baptiste Bourguignon d'Anville, *Éclairissemens géographiques sur la carte de l'Inde* (Paris: Imprimerie Royale, 1753), iii–vi.
75. Edney, "Reconsidering Enlightenment Geography," 186.
76. Jean Baptiste Bourguignon d'Anville, *Carte de l'Inde* (Paris, 1752). See BN, Cartes, GE D 10595 (1–2).
77. Bougainville, *Voyage*, 187.
78. D'Anville, *Éclairissemens*, 148.
79. Jean Baptiste Bourguignon d'Anville, *A Geographical Illustration of the Map of India*, trans. William Herbert (London, 1759).
80. John Green, *Remarks in support of the new chart of North and South America* (London, 1753), 3.
81. Quoted in Cook, "Sea Routes," 123; Cook, "Surveying the Seas," 72–73; Chapuis, *A la mer comme au ciel*, 239; and Howard T. Fry, *Alexander Dalrymple (1737–1808) and the Expansion of British Trade* (Toronto: University of Toronto Press, 1970), 223–24.
82. Charles Noble, *The French and English Marine Regulations Compared* (London, 1755), 3.
83. William Herbert, *A New Directory for the East Indies* (London, 1758).

84. Herbert, *Directory*, dedicatory page.
85. Fisher, *Blueback Charts*, 4.
86. Pedley, *Commerce of Cartography*, 33; and J. B. Harley, "The Bankruptcy of Thomas Jefferys: An Episode in the Economic History of Eighteenth Century Map-making," *Imago Mundi* 20 (1966): 35–37.
87. Thomas Jefferys, *Explanation of the Map on the Seat of War on the Coast of Choramandel* (London, 1754), 6.
88. Jefferys, *Choramandel*, 10.
89. Fry, *Dalrymple*, 222.
90. Johan Splinter Stavorinus, *Voyage to the East-Indies* vol. 2, trans. Samuel Hull Wilcocke (London: G.G. and J. Robinson, 1798), 168–69. Also see Fry, *Dalrymple*, 222.
91. A. R. T. Jonkers, *Earth's Magnetism in the Age of Sail* (Baltimore: Johns Hopkins Press, 2003), 139.
92. Jefferys, *Choramandel*, 36.
93. Edney, "Reconsidering Enlightenment Geography," 186.
94. *The London Magazine* 23 (1754), 440. Note that images are not paginated so that the map appears between pages 440 and 441.
95. "His Majesty's most gracious Speech to Both Houses of Parliament, November 18, 1760," *The Gentleman's Magazine* 30 (1760), 516.
96. See Prakash, "The English East India Company," 8; and Bowen, "No Longer Mere Traders," 25–28.
97. *The London Magazine* 29 (1760), 313.
98. "List of the East India Company's Forts," *The London Magazine* 30 (1761), 344. Map is found on the page following 344.
99. "A brief Account of the Phillippine Islands, with a Chart of the Channel through which the Manila Galleon passes, and a Map of Manilla Island, finely coloured," *The London Magazine* 32 (1763), 292.
100. Marshall, "British in Asia," 487.
101. Chapuis, *A la mer comme au ciel*, 250. A couple of years into the Seven Years' War (1756–1763), d'Après found himself hospitalized in Pondicherry suffering from scurvy. This CIO pilot, who had launched the charting of this region, abandoned the sea; he returned to Lorient in early 1759, and, three years later in 1762, took up the post of inspector at the newly established Company hydrographic office with a pension of 2,400 livres.
102. Jean Baptiste d'Après de Mannevillette, *Mémoire sur la Navigation de France aux Indes* (Paris: Imprimerie Royale, 1768).
103. Jean Baptiste d'Après de Mannevillette, *Instructions sur la navigation des Indes Orientales et de la Chine* (Paris, 1775), xxxiii–iv.
104. McClellan, *Colonial Machine*, 190. D'Après accused Bellin of plagiarizing his *Neptune Oriental*, a view that was upheld by Académie.
105. McClellan, *Colonial Machine*, 190.
106. For information on the life and career of Alexander Dalrymple see Fry, *Dalrymple*; and Cook, "Sea Routes," 127–35.
107. Fry, *Dalrymple*, 9.
108. Fry, *Dalrymple*, 16–35.
109. Fry, *Dalrymple*, 29.
110. Fry, *Dalrymple*, 32.
111. Fry, *Dalrymple*, 35.
112. Fisher, *Blueback Charts*, 10. Also see James Cook, *Captain Cook's Journal . . . 1768–1771*, trans. W. J. L. Wharton (London, 1893), entry 7 September 1770, http://gutenberg.net.au/ebooks/e00043.html#ch9.
113. Fisher, *Blueback Charts*, 10; and Cook, *Captain Cook's Journal*, entry, 7 September 1770.
114. Alexander Dalrymple, *Essay on the Most Commodious Method of Marine Surveying* (London, 1771), 1.
115. Alexander Dalrymple, *General Introduction to the Charts and Memoirs* (London, 1772), viii.

116. Dalrymple, *Charts and Memoirs*, 28.
117. Fry, 223; Alexander Dalrymple, *General Collection of Nautical Publications* (London, 1783); and Alexander Dalrymple, *Chart of the China Seas* (London, 1771), BN, Cartes, GE DD 2987 (7108).
118. Andrew S. Cook, "Alexander Dalrymple's A *Collection of Plans and Ports in the East Indies* (1774-1775): A Preliminary Examination," *Imago Mundi* 33 (1981): 47.
119. Alexander Dalrymple, *A Collection of Plans of Ports in the East Indies* (London, 1775), BL, Maps C. 25. c. 11.
120. Cook, "*Collection of Plans and Ports*," 46–47.
121. Dalrymple, *Collection*, 9.
122. Dalrymple, *Collection*, 13–14.
123. Dalrymple, *Collection*, 15.
124. Fry, *Dalrymple*, 229–30.
125. Cook, "Collection," 51.
126. Cook, "Sea Routes," 130.
127. Fry, *Dalrymple*, 231–32.
128. Fry, *Dalrymple*, 232.
129. Cook, "Sea Routes," 134.
130. Fry, *Dalrymple*, 236.
131. Cook, "Sea Routes," 134.
132. B. E. Kennedy, "Anglo-French Rivalry in Southeast Asia 1763-93: Some Repercussions," *Journal of Southeast Asian Studies* 4, no. 2 (1973): 207.
133. Kennedy, "Southeast Asia," 207, 209, n. 47.
134. Kennedy, "Southeast Asia," 207.
135. Étienne Hulot, *D'Entrecasteaux, 1737–1793* (Paris: Bulletin de la Societé de Géographie, 1894), 22–24.
136. Kennedy, "Southeast Asia," 213. Also see Gilles Palsky, "La Cartographie française des Côtes Cochinchinoises à la Fin du 18e siècle: Jean-Marie Dayot et le 'Pilote de Cochinchine'," *Imago Mundi* 41 (1989): 59–69.
137. Palsky, "Jean-Marie Dayot," 59; Kennedy, "Southeast Asia," 213.
138. Palsky, "Jean-Marie Dayot," 59.
139. Fry, *Dalrymple*, 261.
140. Fry, *Dalrymple*, 261–62.
141. Fisher, *Blueback Charts*, 10.
142. Fisher, *Blueback Charts*, 11.
143. Fisher, *Blueback Charts*, 10–11.
144. Chapuis, *A la mer comme au ciel*, 749.

Conclusion

In the late sixteenth century the relationship between the court of Elizabeth I and Christopher Saxton produced the *Atlas of the Counties of England and Wales* (1579) with the ornate frontispiece of an enthroned Elizabeth. While the work could display for Elizabeth the lands of her dominion, it also provided a view of those distinct territories for the powerful landed elites. As Helgerson has suggested an English audience was disabused of the crown's dynastic primacy as the works that followed displaced the monarch from the central image and joined to it a myriad of other symbols of elites and corporate bodies.[1] About a century later (1682), William Morgan published his survey of London on which the viewer could behold the patronage connection between the late John Ogilby and Charles II. This scene, coupled with the equestrian statues of Charles I and Charles II, along with Morgan's bold statements supporting the king's catholic leanings, presented the viewer with a "blatantly royalist" work,[2] but not at the expense of all the other supportive elite and corporate bodies in and around London. The map is not a celebration of the dominion of Charles II over the city, but an appreciation of the encouragement of those of privilege for the work of Ogilby and Morgan, as expressed through its full of lists of names of influential patrons (catholic and protestant), companies, and important offices. By the end of the century with parliament-centered rule, the image of monarchic dominion over the lands of Britain would be a denial of the British form of government and simply impolitic. The works that followed might have enjoyed royal patronage and would include the royal arms along with those of many other supporters or subscribers to the map or atlas often with separate pages or added borders to include their heraldic shields.

In France, the late sixteenth-century and early seventeenth-century image of Bourbon rule was bold. The atlas of Maurice Bouguereau was not a vision for French nobles to see their lands as distinct and independent but a political manifesto declaring the supremacy of Bourbon rule.[3] The portrait of Henri IV in armor that adorned the opening pages of the atlas was crowned by images of Navarre and France, and in word, the king was praised for his glory in battle, for defeating the hydra of heresy, and for bringing peace and prosperity to France. Henri does not share the stage with his nobles. Years later, Henri's grandson Louis XIV brought Bourbon imagery to its zenith. On Jacques Gomboust's map of Paris (1653), published after the upheaval of the Fronde, noble and social elites could view the chateaus assembled around the map as if in repentant audience with the king. Decades later, after the several wars of Louis XIV stretched the finances and patience of many and after he no longer played a personal role on the battlefield, he was depicted in great repose on Jean Baptiste Nolin's map of France (1693). Dressed in armor, he is seated with his foot crushing the figure of Heresy as the grateful figures of Piety and Justice look on. In the late seventeenth and early eighteenth centuries such imagery was out of place as political and military realities revealed an exhausted France. While Bourbon power remained strong and continued to appear in imagery on maps usually as the fleur de lis, baroque tastes in design had declined, Louis played a less personal/visible role in government, and at his death the personal state would be replaced by the bureaucratic state that he had created (many of whose institutions formed the Colonial Machine). The personal monarch may have slipped from the stage but in France the state emerged with various map making offices to serve the needs of the Bourbon government rather than its *gloire*.

Imagery did not disappear from cartographic productions as county and regional mapping in England and France still displayed for the viewer symbols of authority whether they be the arms of local (secular) elites or ecclesiastical leaders, but with that imagery were often scenes, buildings, local sites of interests even symbols to denote the economic base of the region such as viticulture, mining, orchards, or fishing. Imagery aside, in the eighteenth century, what became important for these county or provincial projects was the ability to carry out a survey on site whether locally directed or directed from a distance. As instrumentation improved, what was required was an infrastructure to support and direct the mapping activity.

In England, as the survey projects (whether by triangulation or not) of Henry Beighton, John Senex, and even Joel Gascoyne revealed, skill was only one part of the equation. The later entrepreneurial mathematicians, engravers, or surveyors who hoped to benefit from the award of the Society of Arts also suffered from the disparate nature of county mapping projects in England. The expense could be nearly career-ending as Thomas Jefferys found out. Subscribers could be found in the noble and gentry classes of the county down to the local bookseller and perhaps a London establishment, but this consumer demand was often not great enough to meet the expense of carrying out the survey from the measurements on the ground to engraving the plates and publishing the work, the last two steps often taking place in London.

In Paris, Delisle or d'Anville would find themselves on the receiving end of local information and surveys for their efforts to map dioceses, provinces, or other administrative units. Unlike the English surveyors or mathematicians, neither Delisle nor d'Anville ever conducted fieldwork. They directed the local efforts with questionnaires and specific instructions for drawing the maps. In addition while both Delisle and d'Anville would sell their works at their business establishments, they did not advertise for subscribers, issue announcements in periodical literature, or offer to fill the map with the armorial bearings of privileged subscribers. They were not as bound to the commercial market as their English counterparts and instead, relied on the traditional social institution of patronage. In both England and France, however, some regional map projects did not see the light of day due to lack of interest (i.e., too few subscribers), death of the map maker (e.g., Delisle for the Languedoc project), or the death of a patron or key supporter.

In France as Delisle and d'Anville were carrying out their cartographic work, the Cassini national mapping project was gradually developing and providing information to the Académie and the Crown among other institutions. By 1745 the first national map was complete and served the needs of royal and provincial civil engineering teams developing public work programs. Nothing of the kind yet existed in England, as county mapping projects were a hard sell. Within the decade, Cassini II and his team were back in the field at the request of Louis XV to produce a work of greater detail for military purposes and, in London, members of the Society of Arts had organized to offer £100 for a successful county map. When the Languedoc project languished after the death of Delisle (1726) and tensions arose between Buache and members of the Society of Montpellier, Cassini de Thury took over the project to ensure the quality and uniformity of the whole. In the second half of the eighteenth century, French provincial officials could contract with Cassini de Thury's engineers to produce maps for their needs just as the officials of Burgundy (who had sought out Delisle in the early eighteenth century) did when they contracted with Joseph-Dominique Seguin.

In England, the lack of infrastructure including a trained body of surveyors made every county map project a tenuous plan that might never see the light of day. In the last quarter of the eighteenth century with Britain at war in their American colonies, the importance of good land surveys to military strategy hit home. In a joint effort with the French to extend the geodetic project to England by establishing the difference in longitude between Greenwich and Paris in the 1780s, the English laid the foundation for the Ordnance Survey that began in the last decade of the eighteenth century just as the military concerns grew over the events of the Revolution in France. The Ordnance Survey would bring systematic mapping to England to serve the strategic needs of the state.

As John Pickles wrote, "maps and mappings precede the territory they 'represent.'"[4] This concept was not lost on the actions and words of English and French officials and map makers in the colonial enterprise, who used claims of authoritative knowledge and a scientific basis to bolster the rhetorical power of their maps. In 1710, Père Bobé expressed dismay that Delisle, the geographer of the Académie, had not accorded more land on his maps to the French. Four years later, Bobé wrote again to Delisle that officials were concerned that he had placed boundaries on his recent world map (1714) with which the court did not agree and which might incite foreign governments. In a different political atmosphere four years later, Delisle published his map, *Louisiane* (1718), based on reports of Le Maire, which prompted a debate over the very generous boundaries. His map defined French colonial possessions that the British decried as evidence of encroachment. The correspondence between colonial officials in North America and the government offices in London revealed long-standing grievances over French encroachment by settlement or alliances with native groups. Colonial officials even warned that the absence of settlers in various regions made these encroachments more dangerous as the French argued that a charter of possession actually required settlement, so they would build forts to establish their presence (however slight) and slowly encroach on the lands that the British had not physically occupied.

Both sides made claims to scientific geography but there was as yet no organized or systematic surveying in these disputed areas. In 1753, Green wrote in dismay over the support enjoyed by French map makers, and in the following year he wrote scathingly about those French geographers who "prostitute their pens" and participate in the "lowest chicanery" all to serve the interests of power.[5] Maps became a tool of impressing one's colonial possession on the land.

A few years prior to the Seven Years' War, John Mitchell published his map of North America (1754) with the sanction of the Board of Trade, based on surveys recently conducted in various areas, and a list of latitude and longitude observations. With this map British officials would deny the French any possessions east of the Mississippi River and again argue that Nova Scotia or Acadia extended from the peninsula to the mainland. There was an internal domestic debate over the limits depicted and the effect the map might have on negotiations of compromise. The chancellor was concerned that this map would send the dispute with France to the breaking point as the boundaries and land claims of the British were bolder than ever before. The map was published with the support of the faction that intended to oust the French from their pretended possessions. These "paper wars" between

Britain and France were only brought to an end by the Treaty of Paris (1763), by which France lost her colonial possessions on mainland North America.

For decades French officials in the naval ministry had collected material from royal vessels and worked to free themselves from a reliance on Dutch charts. Just as government support was important for the execution and maintenance of terrestrial mapping, maritime charts for the surety of navigation required an office to supervise quality and accessibility. As with earlier efforts of support, France was out in front establishing a navy hydrographic office in the first quarter of the eighteenth century. Bellin spent many years organizing and studying the numerous documents collected, and in the 1730s began to produce charts for the royal vessels to cover all the oceans of the world. He continued to contextualize his work within the scientific discourse assuring the ministry of his tireless effort to up-date and improve charts for the navy. He expressed appreciation for the minister's patronage and the abundant source material at his disposal, but he also repeated concerns that ships' journals be turned into the Dépôt and that pilots communicate improvements to be made on his charts. His charts of the Channel and accompanying reports were intended to provide the viewer not only with confidence in the navigational information but a greater sense of authority over English charts sold by the publishing firm of Mount and Page. At the end of the century Britain controlled the Channel but had only just begun to establish an institutional framework to direct and support chart making.

In the East Indies, French pilots remained reliant on Dutch works until the publication by a pilot of the Company. While Bellin had benefited from information provided by the CIO, it was the Company pilot, d'Après, who provided the best charts available in his *Neptune Oriental* published with the approbation of the Académie in 1745. D'Après was an agent of the Colonial Machine, and his *Neptune*, a product of the Machine, was widely recognized as superior to English and Dutch works. It was translated by William Herbert, a purser's clerk of the EIC, in 1758 around the same time that Dalrymple arrived in the East Indies as a secretary of the EIC. These two men, d'Après and Dalrymple, both with piloting experience provided navigators with charts and reports that secured routes and opened up new ones. By 1761, Thornton's *English Pilot* sold by Mount and Page since 1703 had seen its last edition. Cook's remarks (1770) on the disreputable and dangerous chart making practices of navigators who would make things up and publishers who offered their works for sale without evidence of their authoritative standing reflected the long outstanding need for a directing and supportive agency. Even in France where such as agency had existed, a royal decree of 1773 emphasized that chart making was the responsibility of the Dépôt and no one could publish a chart without the formal permission of the ministry.

As the eighteenth century progressed, commercial trading interests required quality charts to compete, and scientific expeditions sent by both

British and French governments encountered new, wondrous, and hazardous phenomena and realized the limitations of current charts and provided new information to improve those that existed. Britain was still without a central agency and the EIC only established its hydrographic office in 1779. War in the American colonies and a new sovereign role for the EIC in India demanded more institutional structure for hydrographic concerns and the Admiralty's hydrographic office was established in 1795.

Much of the story so far has been about the need to establish a supportive and directing set of institutional offices with map making and chart making responsibilities. The government of the personal monarch was long gone and the modern bureaucratic state required cartographic and hydrographic productions for its administrative bodies. Under the centralized government of France, an institutional infrastructure was created that gradually over the course of the eighteenth century made mapping and charting a routine activity and a responsibility of the state. By the last quarter of the eighteenth century royal decrees re-affirmed that those processes required permission from the chancellor or appropriate ministry in an effort to limit any commercial products without a seal of approval. The political upheaval in France at the end of the eighteenth century and extending into the nineteenth century, belied the continued production of maps and charts. The French hydrographic office would emerge out of the era of the Revolution with a deservedly exceptional reputation for charting, but the seas were dominated by the British, and in the nineteenth century the Admiralty charts would become the gold standard for its imperial fleet.

In England, private initiatives launched individual projects to map various counties in the absence of a central controlling body. Only the militaristic concerns of the late eighteenth century prompted the government to establish the Ordnance Survey which began to produce valuable maps in the nineteenth century. For mapping the colonies the Board of Trade and Plantations had long complained of the need for maps and would provide information for privileged and well-connected map makers but was never a supporting or directive body with specific personnel for colonial mapping. The Royal Society and Board of Longitude were supportive of individual scientific ventures such as Halley's charting of the English Channel, but again there was no infrastructure for the routine investigation, analysis, drawing, engraving, and printing of maps or charts. Not surprisingly the first body to come forward with an office for the purpose of collecting and disseminating charts was the British East India Company who was compelled to create the office in 1779 due to pressures of war and its sovereign role in India. The Admiralty would follow with its own hydrographic office in 1795 after collecting documents, reports, journals, etc. for well over a century. It would take several more decades for the Admiralty to establish a core of approved charts which served the British fleet leaving the merchant vessels to be served by the popular blueback charts. By the late eighteenth century London had become the center of cartographic production to serve its growing Imperial Machine, but the story of the maps and charts of this Machine are the topic for another book.

NOTES

1. Richard Helgerson "Writing Empire and Nation," in *The Cambridge Companion to English Literature, 1500–1600*, ed. Arthur F. Kinney (Cambridge: Cambridge University Press, 2000), 325.
2. David I. Bower, "Further Light on Ogilby and Morgan's Map of London (1676)," *Imago Mundi* 65, no. 2 (2013): 284.
3. Hilary Ballon, *The Paris of Henri IV* (Cambridge, MA: MIT Press, 1990), 214.
4. John Pickles, *A History of Spaces: Cartographic Reason, Mapping and the Geo-Coded World* (London: Routledge, 2004), 5.
5. John Green, *Remarks in Support of the New Chart of North and South America* (London, 1753), 3; John Green, *The Conduct of the French with Regard to Nova Scotia* (London, 1754), 2.

Bibliography

Andrews, J. H. "Introduction: Meaning, Knowledge, and Power in the Map Philosophy of J. B. Harley." In *The New Nature of Maps: Essays in the History of Cartography*, edited by J. Paul Laxton, 2–32. Baltimore: Johns Hopkins University Press, 2001.

Andrews, Kenneth R. *Trade, Plunder and Settlement: Maritime Enterprise and the Genesis of the British Empire, 1480–1630*. Cambridge: Cambridge University Press, 1984.

Anville, Jean Baptiste Bourguignon d'. *Eclairissemens Géographiques sur la Carte de l'Inde*. Paris, 1753.

———. *A Geographical Illustration of the Map of India*. Translated by William Herbert. London, 1759.

Anonymous. "Curious Correspondence of De L'Isle the Geographer, as to the Limits of Louisiana, Etc." *Historical Magazine* 3 (1859): 231–33.

Après de Mannevillette, Jean Baptiste d'. *Instructions sur la navigation des Indes Orientales et de la Chine*. Paris, Imprimerie Royale, 1775.

———. *Mémoire sur la Navigation de France aux Indes*. Paris: Imprimerie Royale, 1768.

———. *Le Neptune Oriental, ou Routier General des cotes des Indes Orientales et de la Chine. Enrichi de cartes hydrographiques tant generals que particulieres, pour server di'instruction a la navigation de ces differentes mers*. Paris: Imprimerie Royale, 1745.

———. *Le Nouveau Quartier Anglais, ou description et usage d'un nouvel instrument pour observer la latitude sur mer*. Paris: Imprimerie Royale, 1739.

Aylmer, G. E. "Navy, State, Trade, and Empire." In *The Oxford History of the British Empire: The Origins of Empire*, edited by Nicholas Canny, 467–79. Oxford: Oxford University Press, 1998.

Ballon, Hilary. *The Paris of Henri IV: Architecture and Urbanism*. Cambridge, MA: MIT Press, 1990.

Banks, Kenneth. *Chasing Empire across the Sea: Communications and the State in the French Atlantic, 1713–1763*. Montreal: McGill-Queen's University Press, 2002.

Barber, Peter. "England I: Pageantry, Defense, and Government: Maps at Court to 1550." In *Monarchs, Ministers, and Maps: The Emergence of Cartography as a Tool of Government in Early Modern Europe*, edited by David Buisseret, 26–56. Chicago: University of Chicago Press, 1992.

———. "England II: Monarchs, Ministers, and Maps, 1550–1625." In *Monarchs, Ministers, and Maps: The Emergence of Cartography as a Tool of Government in Early Modern Europe*, edited by David Buisseret, 57–98. Chicago: University of Chicago Press, 1992.

———, ed. *The Map Book*. New York: Walker and Company, 2005.

———. "Was Elizabeth I Interested in Maps—And Did it Matter?" *Transactions of the Royal Historical Society* 14 (2004): 185–98.

Beik, William. *Absolutism and Society in Seventeenth-Century France: State Power and Provincial Aristocracy in Languedoc*. Cambridge: Cambridge University Press, 1985.

Bellin, Jacques Nicolas. *Description Géographique des Debouquemens qui sont au nord de l'isle de Saint Domingue*. Paris: Didot, 1768.

———. *Description Géogrpahique des Isles Antilles possédées par les Anglois*. Paris: Didot, 1758.

———. *Essai Géographique sur les isles Britanniques*. Paris: Didot, 1757.

———. *Observations sur la carte de la Manche*. Paris: Chardon, 1749.

———. "Remarques de M. Bellin, ingenieur de la Marine, sur les Cartes et Plans." In *Histoire et Description de la Nouvelle France*, edited by Père Charlevoix. Paris: Giffart, 1744.

———. *Remarques sur la Carte de l'Amérique Septentrionale*. Paris, 1755.

Blanning, T. C. W. *The Culture of Power and the Power of Culture: Old Regime Europe 1660–1789*. Oxford: Oxford University Press, 2002.

Blond, Stéphane J. L. "The Trudaine Atlas: Government Road Mapping in Eighteenth-Century France." *Imago Mundi* 65, no. 1 (2013): 64–79.

Bluche, François. *Louis XIV*. Translated by Mark Greengrass. New York: Franklin Watts, 1990.

Boislisle, A. M. de, ed. *Mémoires des Intendants sur L'État des Généralités dresses pour l'instruction du duc de Bourgogne*, Book One, *Mémoire de la Généralité de Paris*. Paris: Imprimerie Nationale, 1881. See report on Paris as part of this project and printed in 1881. URL at the BN, http://gallica.bnf.fr/ark:/12148/bpt6k64389722/f484.image.

Bougainville, Louis-Antoine de. *Voyage Round the World*. Translated by John Reinhold Forster. London: J. Nourse, 1772.

Bouguereau, M. *Le Théatre Françoys Tour 1594*. Amsterdam: Theatrum Orbis Terrarum, LTD., 1966.

Bouquet, Simon and Estienne Pasquier. *Bref et sommaire recueil de ce qui a esté faict, [et] de l'ordre tenüe à la ioyeuse [et] triumphante entrée de tres-puuissant . . . prince Charles IX*. . . Paris, 1572.

Bourguet, Marie Noëlle, Christian Licoppe, and H. Otto Sibum, eds. *Instruments, Travel and Science: Itineraries of Precision from the Seventeenth to the Twentieth Century*. New York: Routledge, 2002.

Bousquet-Bressolier, Catherine. "Matthäus Merian's 1615 Map of Paris: Its Structure, Decoration and Message." *Imago Mundi* 58 (2006): 48–69.

Bowen, H. V. "'No Longer Mere Traders': Continuities and Change in the Metropolitan Development of the East India Company, 1600-1834." In *The Worlds of the East India Company*, edited by H. V. Bowen, Margarette Lincoln, and Nigel Rigby, 19–32. Woodbridge, Suffolk: Boydell Press, 2002.

Bower, David I. "Further Light on Ogilby and Morgan's Map of London (1676)." *Imago Mundi* 65, no. 2 (2013): 280–87.

Braddick, Michael J. "The English Government, War, Trade, and Settlement, 1625–1688." In *The Oxford History of the British Empire: The Origins of Empire*, edited by Nicholas Canny, 286–308. Oxford: Oxford University Press, 1998.

Briot, Claude. "L'apport de d'Après de Mannevillette dans les progress de la navigation en Mers des Indes au XVIIIe siècle." In *Autour de d'Après de Mannevillette: Savant Navigateur Havrais au Siècle des Lumières*, edited by Eric Saunier, 33–60. Le Havre: Imprimerie la Petite Presse Le Havre, 2008.

Brown, Lloyd A. *The Story of Maps*. Boston: Little, Brown Company, 1950.

Bruijn, Jaap R. and Femme S. Gaastra. *Ships, Sailors and Spices: East India Companies and their Shipping in the 16th, 17th, and 18th Centuries*. Amsterdam: NEHA, 1993.

Brunelle, Gayle K. "The Images of Empire: Franç̧is I and his Cartographers." In *Princes and Princely Culture 1450–1650*, volume one, edited by M. Gosman, A. MacDonald, and A. Vanderjagt, 81–102. Leiden: Brill, 2003.

Buisseret, David. "Monarchs, Ministers, and Maps in France before the Accession of Louis XIV." In *Monarchs, Ministers, and Maps: The Emergence of Cartography as a Tool of Government in Early Modern Europe*, edited by David Buisseret, 99–123. Chicago: University of Chicago Press, 1992.

Burke, Peter. *The Fabrication of Louis XIV*. New Haven, CT: Yale University, 1992.
Chabert, Joseph Bernard, marquis de. *Voyage fait par ordre du roi en 1750 et 1751 dans l'Amerique septentrionale, pour rectifier les cartes des côtes de l'Acadie, de l'Isle-Royale et de l'isle de Terre-Neuve*. Paris: Imprimerie Royale, 1753.
Chapuis, Olivier. "L'émergence des nouvelles cartes marine: l'oeuvre de Beautemps-Beaupré à la fin du XVIIIème et au début du XIXème siècle." *Imago Mundi* 44 (1992): 90–98.
———. *A la mer comme au ciel: Beautemps-Beaupré & la naissance de l'hydrographie modern (1700–1850)*. Paris: Presses de l'Université de Paris-Sorbonne, 1988.
Charpentier, François. *Discours d'un fidele sujet du roy, touchant l'establissement d'une compagnie françoise pour le commerce des Indes Orientales: adressé à tous les François*. Paris, 1665.
Chevalier, François. "Sur un Manière de Lever la Carte d'un Pays." *Histoire de l'Académie royale des sciences* (1707): 141–45.
Clarke, G. N. G. "Taking Possession: The Cartouche as Cultural Text in Eighteenth-Century American Maps." *Word & Image* 4, no. 2 (1988): 455–74.
Clément, Pierre, ed. *Lettres, instructions et mémoires de Colbert*, vol. 3 pt. 1 & 2, vol. 5. Paris: Imprimerie Impériale, 1861–1873.
Collins, James B. *The State in Early Modern France*. Cambridge: Cambridge University Press, 1995.
Cook, Alan H. *Edmond Halley: Charting the Heavens and the Seas*. Oxford: Oxford University Press, 1998.
———. "Edmond Halley and the Magnetic Field of the Earth." *Notes and Records of the Royal Society of London* 55, no. 3 (2001): 477–81.
Cook, Andrew. "Alexander Dalrymple's 'A Collection of Plans of Paris in the East Indies' (1774–1775): A Preliminary Examination." *Imago Mundi* 33 (1981): 46–64.
———. "Establishing the Sea Routes to India and China: Stages in the Development of Hydrographical Knowledge." In *The Worlds of the East India Company*, edited by H. V. Bowen, Margarette Lincoln, and Nigel Rigby, 119–36. Woodbridge, Suffolk: Boydell Press, 2002.
———. "Surveying the Seas: Establishing the Sea Routes to the East Indies." In *Cartographies of Travel and Navigation*, edited by James R. Akerman, 69–93. Chicago: University of Chicago Press, 2006.
Cook, James. *Captain Cook's Journal . . . 1768–1771*. Translated by W. J. L. Wharton. London, 1893.
Corbett, Margery and Ronald Lightbown. *The Comely Frontispiece: The Emblematic Title-page in England 1550–1660*. London: Routledge & Kegan Paul Ltd., 1979.
Cormack, Lesley B. "Britannia Rules the Waves?: Images of Empire in Elizabethan England." In *Literature, Mapping, and the Politics of Space in Early Modern Britain*, edited by Andrew Gordon and Bernhard Klein, 45–68. Cambridge: Cambridge University Press, 2001.
———. *Charting an Empire: Geography at the English Universities, 1580–1620*. Chicago: University of Chicago Press, 1997.
Cornwall, Henry. *Observations upon Several Voyages to India Out and Home*. London, 1720.
Costa, Shelley. "The *Ladies' Diary*: Gender, Mathematics, and Civil Society in Early-Eighteenth-Century England." *Osiris* 17 (2002): 49–73.
———. "Marketing Mathematics in Early Eighteenth-Century England: Henry Beighton, Certainty, and the Public Sphere." *History of Science* 40 (2002): 211–32.
Crawford, Katherine. *Perilous Performances: Gender and Regency in Early Modern France*. Cambridge: Harvard University Press, 2004.
Crone, G. R. "'The Mariners Mirror' 1588." *The Geographical Journal* 119, no. 4 (1953): 455–58.
Crowley, John E. "'Taken on the Spot': The Visual Appropriation of New France for the Global British Landscape." *The Canadian Historical Review* 86, no. 1 (2005): 1–28.
Cumming, William P. *British Maps of Colonial America*. Chicago: Chicago University Press, 1974.
———. *The Southeast in Early Maps*. Chapel Hill: University of North Carolina Press, 1962.
Dainville, François de. *Cartes Anciennes de l'Eglise de France*. Paris: Librairie Philosophique J. Vrin, 1956.

———. *Cartes Ancienne du Languedoc XVIe-XVIIIe siècles*. Montpellier: Societé languedocienne de géographie, 1961.

———. "Cartes de Bourgogne du XVIIIe siècle." In *Actes du 84e Congrès National des Sociétes Savantes, Dijon, 1959*, 1–19. Paris: Comité des travaux historiques et scientifiques, 1960.

———. "Le Théatre François de M. Bouguereau, 1594. Premier atlas national de France." *Actes du 85e Congrès national des societies savants, Section de géographie*. Paris, 1961.

Dalrymple, Alexander. *Essay on the most commodious method of Marine Surveying*. London, 1771.

———. *General Introduction to the Charts and Memoirs*. London, 1772.

Daniell, F. H. B. and F. Bickley. *Calendar of State Papers, Domestic Series of the Reign of Charles II*, volume 22. London: H. M. Stationary Office, 1921.

Darlington, Ida and James Howgego. *Printed Maps of London, circa 1553–1850*. London: George Philip & Son Limited, 1964.

Dassié, F. *Le Routier des Indes Orientales et Occidentales*. Paris: Jean de la Caille, 1677.

Davies, K. G., ed. *Calendar of State Papers, Colonial Series, America and West Indies*, vol. 44. London: H. M. Stationary Office, 1969.

———, ed. *Calendar of State Papers, Colonial Series, America and West Indies* vol. 45. London: H. M. Stationary Office, 1994.

Dawson, Nelson-Martin. *L'Atelier Delisle: L'Amérique du Nord sur la table à dessin*. Sillery, Québec: Editions du Septentrion, 2000.

Dedijer, Stevan. "The Rainbow Scheme: British Secret Service and Pax Britannica." In *Clio Goes Spying: Eight Essays on History of Intelligence*, edited by Wilhelm Agnell and Bo Huldt. Lund: University of Lund, Sweden, 1983.

Delanglez, Jean. "Franquelin, Mapmaker." *Mid-America* 25 (1943): 29–74.

———. "The Sources of the Delisle Map of America, 1703." *Mid-America* 25 (1943): 275–98.

Delano-Smith, Catherine and Roger J. P. Kain. *English Maps: A History*. Toronto: University of Toronto Press, 1999.

———. "Milieus of Mobility: Itineraries, Route Maps, and Road Maps." In *Cartographies of Travel and Navigation*, edited by James R. Akerman, 16–68. Chicago: University of Chicago Press, 2010.

De Vorsey, Louis Jr. "Maps in Colonial Promotion: James Edward Oglethorpe's Use of Maps in 'Selling' the Georgia Scheme." *Imago Mundi* 38 (1986): 35–45.

Dodge, Martin, Rob Kitchin, and Chris Perkins, eds. *Rethinking Maps: New Frontiers in Cartographic Theory*. New York: Routledge, 2009.

Dugast-Matifeux, Charles. "Notice sur Pierre Garcie-Ferrande et son routier de la mer." *Annales de la Societe Academique de Nantes* 38 (1867): 3–22.

Eaton, Daniel Cady. *A Handbook of Modern French Sculpture*. New York: Dodd, Mead and Company, 1913.

Eccles, W. J. *France in America*, revised edition. East Lansing: Michigan State University Press, 1990.

Edney, Matthew H. "Cartography without 'Progress': Reinterpreting the Nature and Historical Development of Mapmaking." *Cartographica* 30, no. 2 & 3 (1993): 54–68.

———. "John Mitchell's Map of North America (1755): A Study of the Use and Publication of Official Maps in Eighteenth-Century Britain." *Imago Mundi* 60, no. 1 (2008): 63–85.

———. "Reconsidering Enlightenment Geography and Map Making: Reconnaissance, Mapping, Archive." In *Geography and Enlightenment*, edited by David N. Livingstone and Charles W. J. Withers, 165–98. Chicago: University of Chicago Press, 1999.

Elton, G. R. *The Tudor Constitution: Documents and Commentary*. Cambridge: Cambridge University Press, 1982.

Escolar, Marcelo. "Exploration, Cartography and the Modernization of State Power." In *State/Space: A Reader*, edited by Neil Brenner, Bob Jessop, Martin Jones, Gordon MacLeod, 29–52. Oxford: Blackwell Publishing, 2003.

Essenhigh, N. R., Rear Admiral. "The Hydrographic Office: The Next 200 Years." *Naval Review* 83, no. 2 (1995): 90–95

Favier, René. *Les Européens et les Indes Orientales au XVIIIe siècle: Aspects maritimes, commerciaux et coloniaux.* Paris: Ophrys, 1997.
Filliozat, Manonmani. "J.B. D'Après de Mannevillette et la Cartographie de la Nouvelle Route des Indes." *CFC* 175 (2003): 6–16.
———. "D'Après de Mannevillette, Captain and Hydrographer to the French East India Company (1707–1780)." *Indian Journal of History of Science* 29, no. 2 (1994): 329–42.
———. "Le Neptune Oriental: Une Somme de la Cartographie de la Compagnie des Indes." *Cahiers de la Compagnie des Indes* 3 (1998): 21–30.
Fischlin, Daniel. "Political Allegory, Absolutist Ideology, and the 'Rainbow Portrait' of Queen Elizabeth I." *Renaissance Quarterly* 50, no. 1, (1997): 175–206.
Fisher, Susanna. *The Makers of the Blueback Charts: A History of Imray Laurie Norie & Wilson Ltd.* Ithaca, NY: Regatta Press Limited, 2001.
Fordham, H. G. *John Cary, Engraver, Map, Chart and Printseller and Globemaker 1754 to 1835* Cambridge: Cambridge University Press, 1925.
———. "The Listes Générales des Postes de France, 1708-1779, and the Jaillots, géographes ordinaires du roi." *The Library: A Quarterly Review of Bibliography*, 4th series, 3, no. 2 (1923): 115–36.
———. *Les Routes de France.* Geneva: Slatkine-Megariotis Reprints, 1975.
———. "Some Surveys and Maps of the Elizabethan Period Remaining in Manuscript." *The Geographical Journal* 71, no. 1 (1928): 50–60.
———. *Studies in Carto-Bibliography.* Oxford: Clarendon Press, 1914.
Freer, Martha Walker. *History of the Reign of Henry IV, King of France and Navarre*, vol. II. London: Hurst and Blackett, 1860.
Fry, Howard T. *Alexander Dalrymple (1737–1808) and the Expansion of British Trade.* Toronto: University of Toronto Press, 1970.
Galissonière, Roland-Michel Barrin, marquis de La and Etienne de Silhouette, "Memoires des Commissaires du Roi, du 4 Octobre 1751 . . . concernant l'Acadie." In *Memoires des Commissaires du Roi et de ceux de sa Majesté Britannique, sur les Posessions & les Droits respectifs des deux Couronnes en Amérique*, vol. I. Paris: Imprimerie Royale, 1755.
Gascoigne, John. *Science in the Service of Empire: Joseph Banks, The British State and the Uses of Science in the Age of Revolution.* Cambridge: Cambridge University Press, 1998.
Gimson, Basil L. and Percy Russell, *Leicestershire Maps, A Brief Survey.* Leicester: Edgar Backus, 1947.
Giraud, Marcel. *A History of French Louisiana*, Volume One: *The Reign of Louis XIV, 1698–1715.* Translated by Joseph C. Lambert. Revised and corrected by Marcel Giraud. Baton Rouge: Louisiana State University Press, 1974.
———. *Histoire de La Louisiane Française,* Volume Two. Paris: Presses Universitaires de France, 1958.
Gordon, Andrew. "Performing London: the map and the city in ceremony." In *Literature, Mapping, and the Politics of Space in Early Modern Britain,* edited by Andrew Gordon and Bernhard Klein, 69–88. Cambridge: Cambridge University Press, 2001.
Graziani, René. "The 'Rainbow Portrait' of Queen Elizabeth I and its Religious Symbolism." *Journal of the Warburg and Courtauld Institutes* 35 (1972): 247–59.
Green, John. *The Conduct of the French with Regard to Nova Scotia.* London, 1754.
———. *Remarks on the map of North and South America.* London, 1753.
Gronim, Sara Stidstone. "Geography and Persuasion: Maps in British Colonial New York." *The William and Mary Quarterly* 58, no. 2 (2001): 373–402.
Gwyn, Julian. *Frigates and Foremasts: The North American Squadron in Nova Scotia Waters, 1745–1815.* Vancouver: University of British Colombia Press, 2003.
Harding, Richard. *Seapower and Naval Warfare, 1650–1830.* London: University College, London Press, 1999.
Harley, J. B. "The Bankruptcy of Thomas Jefferys: An Episode in the Economic History of Eighteenth Century Map-Making." *Imago Mundi* 20 (1966): 27–48.
———. "Deconstructing the Map." In *The New Nature of Maps: Essays in the History of Cartography*, edited by Paul Laxton, 149–96. Baltimore and London: Johns Hopkins University Press, 2001.

———. "English County Map-Making in the Early Years of the Ordnance Survey: The Map of Surrey by Joseph Lindley and William Crosley." *The Geographical Journal* 132, no. 3 (1966): 372–78.

———. "Maps, Knowledge, and Power." In *The New Nature of Maps: Essays in the History of Cartography*, edited by Paul Laxton, 51–82. Baltimore: Johns Hopkins University Press, 2001.

———. "Power and Legitimation in the English Geographical Atlases of the Eighteenth Century." In *Editing Early and Historical Atlases*, edited by Joan Winearls, 161–204. Toronto: University of Toronto Press, 1995.

———. "The Remapping of England 1750-1800." *Imago Mundi* 19 (1965): 56–67.

———. "Silences and Secrecy: The Hidden Agenda of Cartography in Early Modern Europe." In *The New Nature of Maps: Essays in the History of Cartography*, edited by Paul Laxton, 83–108. Baltimore and London: Johns Hopkins University Press, 2001.

———. "The Society of Arts and the surveys of the English counties 1759–1809." *Journal of the Royal Society of Arts* 112, no. 5090 (1964): 119–24.

Harrison, Stephen, William Kip, Thomas Dekker, and Ben Jonson. *The Arch's of Triumph.* London, 1604.

Harvey, P. D. A. and H. Thorpe. *The Printed Maps of Warwickshire 1576–1900.* Birmingham: University of Birmingham, 1959.

Haudrère, Philippe. *La Compagnie française des Indes au XVIIIe siècle, tome I*, 2nd edition. Paris: Les Indes Savantes, 2005.

———. *Les Compagnies des Indes Orientales: Trois siècles de rencontre entre Orientaux et Occidentaux (1600–1858).* Paris: Editions Desjonqères, 2006.

———. "La Route des Indes au XVIIIème siècle." *Cahiers de la Compagnie des Indes* 3 (1998): 7–14.

———. "The 'Compagnie des Indes' and Maritime Matters (c. 1725–1770)." In *Ships, Sailors and Spices: East India Companies and Their Shipping in the 16th, 17th, and 18th Centuries,* edited by Jaap R. Bruijn and Femme S. Gaastra, 81–97. Amsterdam: NEHA, 1993.

Headlam, Cecil, ed. *Calendar of State Papers, Colonial Series, America and West Indies, 1699.* London: H. M. Stationary Office, 1908.

———. *Calendar of State Papers, Colonial Series, America and West Indies, 1708–1709.* London: H. M. Stationary Office, 1922.

———. *Calendar of State Papers, Colonial Series, America and West Indies, August 1714–December 1715.* London: Her Majesty's Stationery Office, 1928.

———. *Calendar of State Papers, Colonial Series, America and West Indies, August 1717–December 1718.* London: Her Majesty's Stationery Office, 1930.

———. *Calendar of State Papers, Colonial Series, America and West Indies, January 1719 to February 1720.* London: His Majesty's Stationery Office, 1933.

———. *Calendar of State Papers, Colonial Series, America and West Indies, March 1720–December 1721.* London: His Majesty's Stationary Office, 1933.

Heawood, Edward. *English County Maps.* London: Royal Geographical Society, 1932.

Hedin, Thomas F. "The Petite Commande of 1664: Burlesque in the Gardens of Versaille." *Art Bulletin* 83, no. 4 (2001): 651–85.

Helgerson, Richard. *Forms of Nationhood: The Elizabethan Writing of England.* Chicago: University of Chicago Press, 1992.

———. "Writing Empire and Nation." In *The Cambridge Companion to English Literature, 1500-1600*, edited by Arthur F. Kinney, 310–29. Cambridge: Cambridge University Press, 2000.

Hennepin, Louis. *Nouvelle Découverte.* Utrecht, 1697.

Herbert, William. *A New Directory for the East Indies.* London, 1758.

Hewitt, Rachel. *Map of a Nation: A Biography of the Ordnance Survey.* London: Granta Books, 2011.

Hind, Arthur Mayger. *Engraving in England in the Sixteenth & Seventeenth Centuries*, vol. 1. Cambridge: Cambridge University Press, 1952.

Holberton, Edward. *Poetry and the Cromwellian Protectorate: Culture, Politics, and Institutions*. Oxford: Oxford University Press, 2008.

Hornsby, Stephen J. *Surveyors of Empire: Samuel Holland, J.F.W. Des Barres, and the Making of the Atlantic Neptune.* Montreal: McGill-Queen's University Press, 2011.

Howse, Derek. *Greenwich Time and the Discovery of the Longitude.* Oxford: Oxford University Press, 1980.

Hulot, Étienne. *D'Entrecasteaux, 1737–1793.* Paris: Bulletin de la Societé de Géographie, 1894.

Hulse, Clark. *Elizabeth I: Ruler and Legend.* Chicago: University of Illinois Press, 2003.

Jackson, Jack. "The Soupart Map of 1716 and its Influence on Gulf Cartography." *Mapline* nos. 80–81 (Winter 1996/97): 10–11.

Jackson, Jack and Robert S. Weddle. "The Oliván Rebolledo-Saint-Denis Maps of Texas, Louisiana, and New Spain, 1715–1717." In *Mapping Texas and the Gulf Coast: The Contributions of Saint-Denis, Oliván, and Le Maire*, edited by Jack Jackson, Robert Weddle, and Winston Deville, 3–20. College Station: Texas A&M University Press, 1990.

Jacob, Christian. *The Sovereign Map: Theoretical Approaches in Cartography throughout History.* Translated by Tom Conley. Edited by Edward H. Dahl. Chicago: University of Chicago Press, 2006.

Jagger, Graham. "Joseph Moxon, F. R. S. and the Royal Society." *Notes and Records of the Royal Society of London* 49, no. 2 (1995): 193–208.

Jaillot, Alexis Hubert. *Atlas Français.* Paris, 1695

———. *Atlas Nouveau.* Paris, 1681.

Jefferys, Thomas and Thomas Kitchin, *Small English Atlas.* London, 1749.

Jefferys, Thomas. *Le Petit Neptune François: or the French Coasting Pilot.* London, 1761.

———. *Remarks on the French Memorials concerning the Limits of Acadia; printed at the royal printing house of Paris, and distributed by the French ministers at all the foreign courts of Europe with two maps exhibiting the limits one according to the system of the French, as inserted in said memorial; the other conformable to the English rights; as supported by the authority of treaties, continual grants of the French kings, and express passages of the best French authors to which is added an answer to the summary discussion, &c.* London, 1756.

Johnson, Francis R. "Thomas Hood's Inaugural Address." *Journal of the History of Ideas* 3, no. 1 (1942): 94–106

Jonkers, A. R. T. *Earth's Magnetism in the Age of Sail.* Baltimore: Johns Hopkins Press, 2003.

Kennedy, B. E. "Anglo-French Rivalry in Southeast Asia 1763-93: Some Repercussions." *Journal of Southeast Asian Studies* 4, no. 2 (1973): 199–215.

King, James E. *Science and Rationalism in the Government of Louis XIV, 1661–1683.* Baltimore: Johns Hopkins University Press, 1949.

Kitchen, Frank. "John Norden (c. 1547–1625): Estate Surveyor, Topographer, County Mapmaker and Devotional Writer." *Imago Mundi* 49 (1997): 43–61.

Klein, Bernhard. *Maps and the Writing of Space in Early Modern England and Ireland.* Basingstoke, Hampshire: Palgrave, 2001.

Klein, Christopher M. *Maps in Eighteenth-Century British Magazines: A Checklist.* Chicago: the Newberry Library, 1989.

Koeman, C. "Lucas Janszoon Waghenaer: A Sixteenth Century Marine Cartographer." *The Geographical Journal* 131, no. 2 (1965): 202–12.

Konvitz, Josef. *Cartography in France, 1660–1848: Science, Engineering, and Statecraft.* Chicago: University of Chicago Press, 1987.

La Roncière, Monique de. "Manuscript Charts by John Thornton, Hydrographer of the East India Company (1669–1701)." *Imago Mundi* 19 (1965): 46–50.

Laxton, Paul, ed. *The New Nature of Maps: Essays in the History of Cartography* Baltimore: Johns Hopkins University Press, 2001.

Ledward, K. H., ed. *Journals of the Board of Trade and Plantations, Volume 9: January 1750–December 1753.* London: H. M. Stationary Office, 1932.

Licoppe, Christian. "The Project for a Map of Languedoc in Eighteenth-Century France." In *Instruments, Travel and Science: Itineraries of Precision from the Seventeenth to the Twentieth Century*, edited by Marie Noëlle Bourguet, Christian Licoppe, and H. Otto Sibum, 51–74. New York: Routledge, 2002.

Livingstone, David N. and Charles W. J. Withers, eds. *Geography and Revolution.* Chicago and London: University of Chicago Press, 2005.
Lynam, Edward. "Woutneel's Map of the British Isles, 1603." *The Geographical Journal* 82, no. 6 (1933): 536–38.
Lynn, John A. *The Wars of Louis XIV, 1667–1714.* London and New York: Longman Press, 1999.
Macnair, Andrew and Tom Williamson. *William Faden and Norfolk's 18th-Century Landscape.* Oxford: Oxbow Books, 2010.
Maeer, Alistair Simon. "The Cartography of Commerce: The Thames School of Nautical Cartography and England's Seventeenth Century Overseas Expansion." PhD diss., University of Texas at Arlington, 2006.
Marin, Louis. *Portrait of the King.* Translated by Martha M. Houle. Minneapolis: University of Minnesota Press, 1988.
Markham, Clements R. "On the Instruction at Present supplied in this country, in practical astronomy, navigation, route surveying, and mapping." *Proceedings of the Royal Geographical Society and Monthly Record* 4 (1882): 296–314.
Marshall, P. J. Marshall. "The English in Asia to 1700." In *The Oxford History of the British Empire: The Origins of Empire,* edited by Nicholas Canny, 264–85. Oxford: Oxford University Press, 1998.
———. "The British in Asia: Trade to Dominion, 1700–1765." In *The Oxford History of the British Empire: The Eighteenth Century*, edited by P. J. Marshall, 487–507. Oxford: Oxford University Press, 1998.
Mason, A. *Wasn't It Exciting!: A Compilation of the Work of A. Stuart Mason.* London: Royal College of Physicians, 2004.
McClellan III, James E. and François Regouard. "The Colonial Machine: French Science and Colonization in the Ancien Regime." *Osiris*, 2nd series, 15 (2000): 31–50.
———. *The Colonial Machine: French Science and Overseas Expansion in the Old Regime.* Turnhout: Brepols, 2011.
McRae, Andrew. "To Know One's Own: Estate Surveying and the Representation of the Land in Early Modern England." *Huntington Library Quarterly* 56, no. 4 (1993): 333–57.
Middleton, Thomas. *Thomas Middleton: Collected Works.* Edited by Gary Taylor and John Lavagnino. Oxford: Oxford University Press, 2007.
Mildmay, William and Ruvigny de Cosne, "Reply of the English Commissaries . . . to their [French] Memorial of the 4th of October 1751, concerning Nova Scotia or Acadia" In *The Memorials of the English and French Commissaries Concerning the Limits of Nova Scotia or Acadia*, vol. I. London, 1755.
Moll, Herman. *A New Description of England and Wales.* London, 1724.
Monahan, W. Gregory. "Tyrant of Languedoc? Nicolas de Lamoignon de Basville in Public and in Private." *Proceedings of the Western Society for French History* 37 (2009): 21–34.
Morgan, Victor. "The Cartographic Image of 'The Country' in Early Modern England." *Transactions of the Royal Historical Society* 29 (1979): 129–54.
Neumann, Bernard de. "A Brief History of the Royal Hospital School." In *Mariners: The Website of the Mariners Mailing List* http://www.mariners-l.co.uk/GreenwichRoyal.html. Accessed: 17 August 2013.
Noble, Charles. *The French and English Marine Regulations Compared.* London, 1755.
Ogilby, John. *Britannia.* London, 1675.
———. *Britannia*, facsimile edition. Amsterdam: Theatrum Orbis Terrarum LTD, 1970.
———. *The Entertainment of His Most Excellent Majestie Charles II*, 2nd ed. London, 1662.
———. *The Relation of his Majesties entertainment passing through the City of London.* London, 1661.
Palsky, Gilles. "La Cartographie française des Côtes cochinchinoises à la Fin du 18e siècle: Jean-Marie Dayot et le 'Pilote de Cochinchine'." *Imago Mundi* 41 (1989): 59–69.
Passerat, C. *Etude sur les cartes des côte de Poitou et de Saintonge antérieures aux levés du XIXe siècle.* Thèse pour le doctorat présentée à la Faculté des lettres de l'Université de Paris 1910.

Pastoureau, Mireille. "Les Atlas imprimés en France avant 1700." *Imago Mundi* 32 (1980): 45–72.

———. *Les Atlas Français XVIe-XVIIe siècles* Paris: Bibliothèque Nationale, 1984.

———. "Jacques-Nicolas Bellin, French Hydrographer, and the Royal Society in the Eighteenth Century." *Yale University Library Gazette*, 65–69.

———. *Nicolas Sanson d'Abbeville: Atlas du Monde 1665* Paris: Sand & Conti, 1988.

Pease, Theodore Calvin. *Anglo-French Boundary Disputes in the West, 1749–1763.* Springfield: University of Illinois, 1936.

Peck, Linda Levy. *Consuming Splendor: Society and Culture in Seventeenth-Century England.* Cambridge: Cambridge University Press, 2005.

Pedley, Mary Sponberg. *Bel et Utile: The Work of the Robert de Vaugondy Family of Mapmakers.* Tring: Map Collector Publications, 1992.

———. *The Commerce of Cartography: Making and Marketing Maps in Eighteenth-Century France and England.* Chicago: University of Chicago Press, 2005.

———. "'Commode, complet, uniforme, et suivi': Problems of Atlas Editing in Enlightenment France." In *Editing Early and Historical Atlases*, edited by Joan Winearls, 83–108. Toronto: Toronto University Press, 1995.

———. "Map Wars: The Role of Maps in the Nova Scotia/Acadia Boundary Dispute of 1750." *Imago Mundi* 50 (1998): 96–104.

———. "Maps, War, and Commerce: Business Correspondence with the London Map Firm of Thomas Jefferys and William Faden." *Imago Mundi* 48 (1996): 161–73.

Pelletier, Monique. "National and Regional Mapping in France to about 1650." In *History of Cartography: Cartography in the European Renaissance*, Vol. 3, Part 2, edited by David Woodward, 1480–503. Chicago: University of Chicago Press, 2007.

Pennington, Richard. *A Descriptive Catalogue of the Etched Work of Wenceslaus Hollar, 1607–1677.* Cambridge: Cambridge University Press, 1982.

Petto, Christine Marie. *When France was King of Cartography: The Patronage and Production of Maps in Early Modern France.* Lanham, MD: Lexington Books, 2007.

Pickles, John. *A History of Space: Cartographic Reason, Mapping and the Geo-coded World.* London and New York: Routledge, 2004.

Pilgrim, Donald G. "The Colbert-Seignelay Naval Reforms and the Beginniing of the War of the League of Augsburg." *French Historical Studies* 9 (1975): 235–62.

Pluchon, Pierre. *Histoire de la Colonisation Française.* Paris: Libraire Arthème Fayard, 1991.

Postlewahyt, Malachy. *Britain's Commercial Interests Explained and Improved*, vol. 1. London, 1757.

Prakash, Om. "The English East India Company and India." In *The Worlds of the East India Company* edited by H. V. Bowen, Margarette Lincoln and Nigel Rigby, 1–18. Woodbridge, Suffolk: Boydell Press, 2002.

Pritchard, James. *Anatomy of a Naval Disaster: The 1746 French Expedition to North America.* Montreal: McGill-Queen's University Press, 1995.

———. *In Search of Empire: The French in the Americas, 1670–1730.* Cambridge: Cambridge University Press, 2004.

———. *Louis XV's Navy 1748–1762: A Study of Organization and Administration.* Montreal: McGill-Queen's University Press, 1987.

Quinn, David B., ed. *Early Maryland in a Wider World.* Detroit, MI: Wayne State University Press, 1982.

———. *England and the Discovery of America, 1481–1620.* New York: Alfred A. Knopf, 1973.

———. *Explorers and Colonies: America, 1500–1625.* London: The Hambledon Press, 1990.

Ramsey, Ann W. "The Ritual Meaning of Henri IV's 1594 Parisian Entry." In *French Ceremonial Entries in the Sixteenth Century: Event, Image, Text,* edited by Nicolas Russell and Hélène Visentin, 189-206. Toronto: University of Toronto Press, 2007.

Ravenhill, Mary R. and Margery M. Rowe, ed. *Maps of Georgian Devon.* Exeter: Short Run Press, 2002.

Ravenhill, W. "Joel Gascoyne, a Pioneer of Large-Scale County Mapping." *Imago Mundi* 26 (1972): 60–70.

Reinhartz, Dennis. *The Cartographer and the Literati: Herman Moll and his Intellectual Circle.* New York: Edwin Mellon Press, 1997.

———. "Herman Moll, Geographer: An Early Eighteenth-Century European View of the American Southwest." In *The Mapping of the American Southwest*, ed. Dennis Reinhartz and Charles C. Colley, 18–36. College Station: Texas A&M University, 1987.

Ritchie, G. S. "Captain Cook's Influence on Hydrographic Surveying." *Pacific Studies* 1(1978): 78–95.

Robinson, A. H. W. "Marine Surveying in Britain during the Seventeenth and Eighteenth Centuries." *The Geographical Journal* 123, no. 4 (1957): 449–56.

Roland, François. *Alexis-Hubert Jaillot, géographe du roi Louis XIV (1632–1712).* Besançon: Imprimerie Jacques et Demontrond, 1919.

Rose, Susan. "Mathematics and the Art of Navigation: The Advance of Scientific Seamanship in Elizabethan England." *Transactions of the Royal Historical Society* 14 (2004): 175–84.

Roy, William. "An Account of the Measurement of a Base on Hounslow-Heath." *Philosophical Transactions of the Royal Society* 75 (1785): 385–480.

Sandman, Alison and Eric H. Ash. "Trading Expertise: Sebastian Cabot between Spain and England." *Renaissance Quarterly* 57, no. 3 (2004): 813–46.

Saunders, Stewart. "Public Administration and the Library of Jean-Baptiste Colbert." *Libraries & Culture* 26, no. 2 (1991): 283–300.

Saunier, Eric. Ed. *Autour de d'Après de Mannevillette: Savant Navigateur Havrais au Siècle des Lumières.* Le Havre: Imprimerie la Petite Presse Le Havre, 2008.

Schmidt, Benjamin. "Mapping an Empire: Cartographic and Colonial Rivalry in Seventeenth-Century Dutch and English North America." *The William and Mary Quarterly* 54, no. 3 (1997): 549–78.

Schutz, John. *William Shirley, King's Governor of Massachusetts.* Chapel Hill: University of North Carolina Press, 1961.

Seller, John. *The English Pilot, First Book.* London, 1671.

Senex, John. *An Actual Survey of all the Principal Roads of England and Wales; described by one hundred maps form copper plates on which are delineated all the cities, towns, villages, churches, houses, and places of note throughout each road. As also directions to the curious traveller what is worth observing throughout his journey.* London, 1719.

———. *A New and General Atlas* London, 1721.

Sharpe, Kevin. *Politics and Ideas in Early Stuart England: Essays and Studies.* London and New York: Pinter Publishers, 1989.

———. "Representations and Negotiations: Texts, Images, and Authority in Early Modern England." *The Historical Journal* 42, no. 3 (1999): 870–89.

Shirley, William and William Mildmay. "A Memorial Concerning the Limits of Nova Scotia or Acadia, . . . January 11, 1751." In *The Memorials of the English and French Commissaries Concerning the Limits of Nova Scotia or Acadia*, vol. I. London: 1755.

———. "Memorial of His Majesty's Commissaries . . . relative to the Island of St. Lucia." In *The Memorials of the English and French Commissaries concerning St. Lucia.* London, 1755.

Silhouette, Etienne de. "Second Mémoire des Commissaires François, sur l'isle de Sainte-Lucie, du 4 Octobre 1754." In *Memorials of the English and French Commissaries concerning St. Lucia.* London, 1755.

Skelton, R. A. *Saxton's Survey of England and Wales with a Facsimile of Saxton's Wall Map of 1583.* Amsterdam: Israel, 1974.

Smith, Thomas R. "Manuscript and Printed Sea Charts in Seventeenth-Century London: The Case of the Thames School." In *The Compleat Plattmaker,* edited by Norman J.W. Thrower, 45–100. Berkeley: University of California Press, 1978.

Stavorinus, Johan Splinter. *Voyage to the East-Indies,* vol. 2. Translated by Samuel Hull Wilcocke. London: G. G. and J. Robinson, 1798.

Stewart, Mart A. "William Gerard de Brahm's 1757 Map of South Carolina and Georgia." *Environmental History* 16, no. 3 (2011), 524–35.

Strong, Roy. *The Tudor and Stuart Monarchy: Pageantry, Painting, Iconography.* Vol. II. Rochester, NY: Boydell Press, 1995.

Stroup, Alice. *Royal Funding of the Parisian Académie Royales des Sciences during the 1690s.* Philadelphia: The American Philosophical Society, 1987.

Tanner, J. R., ed. *Samuel Pepys's Naval Minutes.* London: Naval Records Society, 1926.

Taylor, E. G. R., ed., *A Brief summe of geographie by Roger Barlow* London: Hakluyt Society, 1932.

———. "Roger Barlow: A New Chapter in Early Tudor Geography." *The Geographical Journal* 74, no. 2 (1929): 157–66.

———. "A Sixteenth-Century Ms. Navigating Manual in the Society's Library." *The Geographical Journal* 78, no. 4 (1931): 346–52.

Thornton, John. *The English Pilot, the Third Book on Oriental Navigation.* London, 1703.

Towner, John. "The Grand Tour: Sources and a Methodology for an Historical Study of Tourism." *Tourism Management* (September 1984): 215–22.

Tyacke, Sarah. "Chartmaking in England and Its Context, 1500–1660." In *The History of Cartography, Volume Three, Cartography in the European Renaissance, Part I*, edited by David Woodward, 1722–1753. Chicago: University of Chicago Press, 2007.

———. "Geography is Better than Divinity: The Practitioners' Story." Sandars Lectures: Conversations with Maps, Cambridge University Library, 6 March 2007. http://www.lib.cam.ac.uk/sandars/tyacke2.pdf. Accessed 7 June 2014.

———. *London Map-sellers 1660–1720: A Collection of Advertisements for Maps Placed in the London Gazette 1668–1719 with Biographical Notes of the Map Sellers.* Tring: Map Collector, 1978.

Tyacke, Sarah and John Huddy. *Christopher Saxton and Tudor Map-Making.* London: British Library, 1980.

Underhill, Arthur. *A Summary of the Law of Torts*, 5th edition. London: Butterworths, 1889.

Vorsey, Louis De. *The Georgia-South Carolina Boundary: A Problem in Historical Geography.* Athens: University of Georgia Press, 1982.

Waghenaer, Lucas Jansz. *The Mariners Mirrour London 1588.* Translated by Anthony Ashley, facsimile edition. Amsterdam: Theatrum Orbis Terrarum LTD., 1966.

Waghenaer, Lucas Jansz. *Thresoor der Zeevaert Leyden 1592*, facsimile edition. Amsterdam: Theatrum Orbis Terrarum LTD., 1965.

Wallis, Helen M. "Geographie is Better than Divinitie. Maps, Globes, and Geography in the Days of Samuel Pepys." In *The Compleat Plattmaker: Essays on Chart, Map, and Globe Making in England in the seventeenth and eighteenth centuries*, edited by Norman J. W. Thrower, 1–43. Berkeley: University of California Press, 1978.

———, ed. *Jean Rotz, The Maps and Text of the Boke of Idrography.* Oxford: Roxburghe Club, 1981.

Walters, G. "Themes in the Large Scale Mapping of Wales in the 18th century." *The Cartographic Journal* 5, no. 2 (1968): 135–46.

Waters, David W. "English Navigational Books, Charts Globes Printed Down to 1600." *Revista da Universidade de Coimbra* 33 (1985): 237–57.

———. *The Rutters of the Sea: The Sailing Directions of Pierre Garcie.* New Haven, CT: Yale University Press, 1967.

Withers, Charles W. J. *Geography, Science and National Identity: Scotland Since 1520.* Cambridge: Cambridge University Press, 2001.

———. *Placing the Enlightenment: Thinking Geographically about the Age of Reason.* Chicago: University of Chicago Press, 2007.

Worms, Laurence. "The Maturing of British Commercial Cartography: William Faden (1749–1836) and the Map Trade." *The Cartographic Journal* 41, no. 1 (2004): 5–11.

———. "The Search for John Senex F.R.S.: An Aspect of the Early Eighteenth-Century Book Trade." Paper delivered to the Bibliographical Society, 18 January 2000. http://laurenceworms.wordpress.com/2014/01/30/the-search-for-john-senex-f-r-s-an-aspect-of-the-early-eighteenth-century-book-trade/. Accessed 30 May 2014.

———. "Thomas Jefferys (1719–1771): Beginning the World Afresh." *Mapforum* 3 (2004): 20–29.

Index

Académie Royale des Sciences (Royal Academy of Sciences), xi, xii, xvi, 45, 49, 52, 53, 54–56, 58, 60, 80, 87, 96, 107, 108, 110, 124, 126, 171, 172, 192n104, 197, 198, 199
Acadie (Acadia), xix, 126, 137, 138, 139, 140, 141, 142, 143, 144, 145, 147, 148, 150, 151, 155n65, 157n96, 157n97, 158n104, 158n106, 198. *See also* Nova Scotia
Admiralty, xvi, xx, 80, 91, 92, 97–98, 100, 101, 108, 110, 112, 113, 117n66, 124, 152, 162, 165, 186–187, 187, 200; Hydrographic Office, 80, 110, 113
Alais, diocese of, 52, 73n35. *See also* Nismes, diocese of; Nolin, Jean Baptiste
Anglo-French Boundary Commission(ers), 138, 139, 141, 142, 143, 151, 156n81; Bedford, John Russell, duke of, 139, 156n81; Mildmay, William, 148, 156n81; Shirley, William, 139, 145, 148, 156n81. *See also* La Galissonière, Roland-Michel Barrin, marquis de
Anville, Jean Baptiste Bourguignon d', 45, 61–63, 71n3, 75n88, 124, 135, 146, 147, 148, 159n123, 174–176, 177, 178, 183, 197
Apian, Peter, 5; Albrecht V, 5; survey of Bavaria (1568), 5
Argenson, Marc-Pierre de Voyer de Paulmy, comte d', 144, 145, 158n106
Ashley, Anthony, 86, 91, 110; *The Mariners' Mirrour* (1587), 86. *See also* Waghenaer, Lucas Janszoon
Atlantic Ocean, 81, 93, 97, 99, 103, 106, 135, 142, 168, 179; coastline, 81, 83, 84, 95, 101, 102, 152, 159n129. *See also* Ponant (Atlantic fleet)

Baconian science, 87, 162
Basville, Nicolas de Lamoignan de, 50, 51, 52
Batavia, 164, 177, 178
Beachy Head, battle of, 94
Beautemps-Beaupré, Charles-François, 80, 113, 186, 187. *See also Dépôt des Cartes*
Beauvilliers, Paul de, duc de Saint-Aignan, 50
Beighton, Henry, 59–60, 61, 75n74, 196; *Ladies Diary*, 59–60
Bellin, Jacques Nicolas, 80, 96, 98–100, 101, 102–104, 107–108, 109, 110, 112, 119n115, 119n116, 124, 137–138, 142, 143, 144, 146, 147, 148, 150, 151, 152, 157n96, 158n109, 169, 170, 171, 174, 180, 189n19, 192n104, 199. *See also Dépôt des Cartes*
Bennett, John, 110, 111, 121n154, 184. *See also* Sayer, Robert

Béveziers, battle of. *See* Beachy Head, battle of
Béziers, diocese of, 54; Rousset, Louis-Charles des Alris de (bishop of Béziers), 54. *See also* Delisle, Guillaume; Gautier, Hubert
Blois, diocese of, 63; Caumartin, Jean François Paul Le Fèvre de (bishop of Blois), 62, 63. *See also* Anville, Jean Baptiste Bourguignon d'
bluebacks, 187, 201
Board of Longitude, xvi, 100, 200
Board of Trade and Plantations, xii, xvi, xix, 124, 125, 128, 130, 131, 132, 133, 134, 140, 143, 144, 148, 151, 159n123, 177, 198, 200; Council of Trade and Plantations, 123
Bobé, Père Jean. *See* Delisle, Guillaume
Bocage, Georges Boissaye du, 90, 107; *Petit Flambeau de la Mer*, 107
Bombay, 163, 166, 167, 178, 180, 183; Catherine of Braganza's dowry, 163
Bougainville, Louis-Antoine de, 173, 175, 182
Bouguereau, Maurice, 10, 11, 13, 15, 32, 195; *Le Théatre Françoys* (1594), 10, 13
Bourne, William, 83; *The Regiment for the Sea* (1574), 83
Brest, 95, 107, 110, 118n96; Académie de Marine, 110
Bristol Channel, 67, 97, 118n88
Britannia (1675). *See* Ogilby, John
British Channel. *See* English Channel
British East India Company, xix, xx, 112, 162, 200
Buache, Philippe, 64, 197. *See also* Delisle, Guillaume; *Dépôt des Cartes*
Burghley, William Cecil, Lord. *See* Cecil, William
Burgundy, Duchy of, 55, 56, 65, 67; mapping of, 55, 57, 65, 67; Estates of, 64, 65; revenue officials, 55, 197. *See also* Delisle, Guillaume; Seguin, Joseph-Dominique
Burnet, William (Governor of New York), 130, 132, 133. *See also* encroachments

Cabot, Sebastian, 81, 114n5, 114n12
Camden, William, 8; *Britannia* (1607), 8
Canada, 84, 108, 127, 128, 130, 131, 133, 137, 138–139, 142, 143, 144, 147, 148, 155n65, 157n96, 159n129. *See also* *Nouvelle France*
Cape of Good Hope, 166, 167, 168, 172, 173
Capell, Arthur, earl of Essex, 47, 48. *See also* Essex, county of
Carolina(s), 123, 127–128, 130, 131, 132, 134, 135, 136
Cartier, Jacques, 84
cartography/cartographic, xii–xiv, xv–xix, xx, 2, 4, 9, 14, 21, 24, 38, 45, 46–47, 48, 50, 52, 53, 54, 64, 70, 77n129, 82, 83, 86, 97, 124, 125–126, 127, 130, 136, 137, 143, 144, 148, 152, 174–175, 176, 187, 196, 197, 200, 201; cartographic modes, xiv, xv. *See also* map making
Cassel, battle of, 32
Cassini, Jacques (Cassini II), 60, 63, 117n70, 197
Cassini, Jean Dominique (Cassini I), 49, 52, 60, 90, 93, 126, 127
Cassini, Jean Dominique (Cassini IV), 70
Cassini de Thury, César François (Cassini III), 63–65, 70, 107, 197
Cassini project to map France, xiv, 46, 54, 56, 61, 63, 117n70, 197. *See also* national mapping project (France)
Catholic League, 10, 13, 14; Guise, Henri de, 10; Guise, Louis, cardinal, 10
Cave, Edward, 104, 137; *See also* *Gentleman's Magazine*
Cavendish, Thomas, 84
Cecil, William, Lord Burghley, 5, 21, 83; Cecil, Robert, Earl of Salisbury (son of William Cecil), 23
Chabert, Joseph-Bernard de Cogolin, marquis de, 138, 139, 147, 149, 152
Champlain, Samuel de, 84
Channel Islands, 103, 105, 119n115, 119n116
Charles I, king of England, 2, 17, 30, 91, 195; Henrietta-Marie, widow of Charles I, 17
Charles II, king of England, 25, 26, 28, 30, 31, 35, 37, 47, 89, 92, 117n66, 132,

163, 195
Charles IX, king of France, 9, 130; Elisabeth, Archduchess of Austria (wife of Charles IX), 9
Charlevoix, Pierre-François-Xavier de, 136, 137, 143
Chevalier, François, 53, 60, 61, 62
China, 168, 172, 179, 180, 181, 184, 185; China Seas, xx, 169, 173, 182, 183, 185
Choiseul, Etienne-François, duc de, 105, 106, 108
Christ's Hospital, school of navigation, 92, 118n85
City of London, 22, 24, 30, 47, 84, 92
Civil War (English), 87
Clarke, George (lieutenant governor of New York), 133, 134; Newcastle, Thomas Pelham Holles, Duke of, 133, 139, 156n81. *See also* encroachments
Clerville, Louis-Nicolas de, 89, 116n45
Coasting Pilot (1693). *See* Collins, Greenville
Cochin China (Vietnam), 180, 185–186, 187
Colbert, Jean Baptiste, xi, 31, 49, 50, 51, 53, 72n28, 87–88, 89–90, 94, 116n45, 164, 165, 189n19
Colbert, Jean-Baptiste Antoine, marquis de Seignelay, 89, 94, 116n45, 116n49
Colbert de Saint Pouange, Jean Baptiste Michel, archbishop of Toulouse, 50, 51
Colbert du Terron, Charles-Jean, 87, 88, 89
Collins, Greenville, xviii, 80, 93, 96, 100, 101, 103, 105, 117n66; *(Great Britain's) Coasting Pilot* (1693), xviii, 79, 93, 96, 100, 101, 103, 105
Colonial Machine (French), xvi, xvii, xviii, xix, xx, 37, 38, 45, 50, 90, 92, 96, 98, 114, 124, 138, 147, 162, 164, 169, 171, 174, 187, 190n51, 196, 199
colonial possessions, xix, 123, 125, 151, 152, 177, 198, 199
Compagnie des Indes Orientale, xvi, xix, 129, 161, 164, 165, 167, 169, 170–171, 172, 174–175, 176, 180, 189n14, 189n19, 192n101, 199; Caron, François, 164; Charpentier, François, 164. *See also* French East India Company
Condé, Prince of, 17–18, 18, 20; Condé, family, 55, 65. *See also* Fronde(s)
Conway, Thomas, comte de Conway, 186
Cook, Captain James, 79, 80, 111, 112, 181, 199
Cook, James, 111
Copland, Robert, 81; *The Rutter of the See* (1528), 81. *See also* Garcie, Pierre
Cornwall, county of, 23, 56–58; Bodville, Charles, earl of Radnor, 57. *See also* Gascoyne, Joel
Cornwall, H. (Captain), 167, 175
Coromandel Coast, 175, 178, 179, 180
Cortés, Martin, 83; *Art of Navigation* (1561), 83; Eden, Richard, 83
Crozat, Antoine, 129, 154n26. *See also* Delisle, Guillaume

Dalrymple, Alexander, xx, 180, 180–182, 183–185, 186–187, 199
Dampier, William, 133, 167
D'Après de Mannevillette, Jean-Baptiste-Nicolas-Denis, xx, 161, 171–174, 175, 176, 179, 180, 182, 183, 184, 190n53, 192n101, 192n104, 199
Dassié, F., 167, 168
Dee, John, 4, 6, 84
Degaulle, Jean-Baptiste, 80, 110, 113, 120n148
Delisle, Claude (father of Guillaume), 54, 126, 127
Delisle, Guillaume, xix, 45, 52, 54–56, 57, 61, 64, 73n48, 124, 126–127, 128–129, 132, 145, 150, 153n10, 153n15, 171, 197; atlas of Languedoc project, 55, 61, 74n58, 197; *Carte de la Louisiane et du cours du Mississipi* (1718), xix, 125, 130, 131, 133, 134, 198; correspondence with Père Jean Bobé, 128, 129, 150, 154n22, 198; map of duchy of Burgundy, 55, 57, 64; *premier géographe du roi*, 45, 54, 71n3, 171
Dépôt des Cartes, Plans et Journaux de la Marine (Dépôt or Hydrographic Office of the Navy), xvi, xviii, 80, 96, 98, 100, 101, 109, 110, 112–113, 119n116, 137, 138, 169–171, 173, 179, 180, 185, 187, 189n19, 199; *Archives de la Marine*, 98, 116n49, 189n19

Deptford Strand, 83, 92, 114n12
Des Barres, J. F. W., 152
Devon, county of, 58, 66, 67, 68; Courtenay, Lord, earl of Devon, 67. *See also* Donn, Benjamin
Dieppe, 82, 107; Denis, Guillaume, 88, 89; hydrographic school, 88; school of cartography, 82
Donn, Benjamin, 66–68
Drake, Sir Francis, 6, 84, 86
Drayton, Michael, 8; *Poly-Olbion* (1612), 8
Dunkirk, 6, 95, 104, 107
Dutch, xiv, xx, 85, 88–89, 91, 93, 94, 95, 96, 99, 129, 161, 162, 163, 164, 165, 166, 167, 169, 172, 177, 181, 184, 185, 186; works, xviii, 79, 81, 86, 87, 93, 109, 116n54, 117n66, 161, 165, 166, 168, 169, 170, 172, 173, 178, 181, 185, 191n61, 199
Dutch Protestants, 3, 4, 84
Dutch War, 31, 32, 43n120, 90
Dutch East India Company, 161, 164, 168

Elizabeth I, 3–5, 6–8, 10, 21, 23, 24, 37, 84, 85, 86, 92, 114n12, 163, 195; *The Armada Portrait* (1588), 3; cult of, 4, 5; Elizabethan, xvi, 23, 26, 84; *The Ditchley Portrait* (1592), 3; Privy Council, 5, 21, 84, 85, 86, 92, 163; *The Rainbow Portrait* (c. 1600), 3
Enciso, Martin Fernandez de, 81; Barlow, Roger, 81, 82, 114n5; *Suma de Geographia* (1519), 81
encroachments (territorial), xix, 104, 107, 123–124, 125, 126, 127, 128, 130–131, 132, 133, 135, 136, 137, 138, 138–140, 142, 144, 145, 146, 147, 151, 154n18, 155n44, 198; paper encroachments, 124, 126, 130, 138, 142, 144
England, invasions of (descents), 5, 70, 71, 79, 94, 95, 99, 105, 106, 108, 109, 109–110, 119n115, 120n143; Guines, Adrien-Louis de Bonnières, comte de, 109, 120n143
English Channel, xviii, 9, 71, 79–80, 81, 86, 88, 89, 93–94, 95, 96, 97, 99, 100, 101, 103, 105, 106, 108, 109, 110, 111, 118n88, 118n96, 123, 168, 169, 200; chart of, xviii, 99, 100, 102, 105, 107, 108, 110, 113, 167, 199
English East India Company, xii, 84, 163–164, 165, 166, 168, 169, 176, 177, 179, 180, 181, 182, 183–185, 187, 188n1, 188n4, 189n12, 189n19, 199, 200. *See also* British East India Company
Enlightenment period, xiv, 38, 46, 62, 64, 126, 165; encyclopedism, xv, 46, 62, 87, 174, 176; enlightened, xvii, 37, 188; epistemology, 98, 162; ideology, 46, 62, 63, 82, 98, 162, 170, 174, 176, 178
Entrecasteaux, Joseph-Antoine d', 185, 186
Essex, county of, 47, 68, 69; André, Peter, 68; Chapman, John, 68, 69; Ogilby, John, 47, 68; Morgan, William, 47; Sparrow, Thomas, 68
exegetical method, xiv, xvii, 45, 126

Finé, Oronce, 9, 83
France, invasions of (descents), 79, 95, 102, 103, 116n48
François I, king of France, 9, 82, 83; Collège Royale, 9, 83
French East India Company, xix, 185
French Revolution, 65, 70, 110, 186, 198, 200
Fréron, Élie Catherine, 65
Frobischer, Martin, 84
Fronde(s), 20, 87, 196; Frondeurs, 18, 21; Parlementary Fronde, 17; Princely Fronde, 17, 18
Fulvy, Jean-Louis Henri Orry de, 172, 191n61

Gallia, 10, 11
Garcie, Pierre, 81, 83, 88, 161; *Le grant routtier* (1520), 81; Ymbert, Pierre, 81
Gascoyne, Joel, 56–58, 66, 68, 74n63, 165, 196
Gautier, Hubert, 50–51, 53, 54
Gentleman's Magazine, xix, 68, 104, 104–106, 107, 126, 136, 137, 139–140, 178
gentry, xii, xvi, 8, 9, 37, 47, 48, 57, 59, 69, 75n83, 77n131, 91, 92, 107, 108, 196

geodesy, xiv; geodetic project, 46, 70, 90, 197
geographical archive, 162, 165, 170, 183
géographe de cabinet (geographers of the study), xiv, 45, 48, 63
géographe du roi (royal geographers), xvi, 43n113, 43n116, 48, 54, 71n3. *See also* *premier géographe du roi*
George II, king of Britain, 58
George III, king of Britain, 66, 80, 113, 178
Georgia, colony of, xix, 135–136; Martyn, Benjamin, 135, 136; Oglethorpe, James Edward, 135
Gheeraerts, Marcus, 3, 4; *The Ditchley Portrait* (1592), 3; *The Rainbow Portrait* (c. 1600), 3
gloire, xi, 4, 31, 35, 37, 38, 41n45, 43n116, 49, 50, 87, 164, 191n61, 196
Glorious Revolution, 2, 28, 31, 37, 93, 163, 166
Gomboust, Jacques, 18, 19, 20, 196
Goos, Pieter, 88, 91, 165, 167, 170, 173, 174, 175; *De Lichtende Columne ofte Zee-Spiegel* (1650), 88; *De Zee-Atlas ofte Water-Wereld* (1666), 88
Gower, George, *The Armada Portrait* (1588), 3
Great Fire (of London), 26, 27, 47
Green, John, 66, 71, 124, 125, 131, 145, 146–147, 148, 152, 176, 198. *See also* Mead, Braddock
Grenier, Jacques-Raymond, 182
Guerard, Nicolas, 55
Guérin, Gilles, *Louis XIV Terrassant La Fronde* (1654), 18
Gunter, Edmund, 92

Hakluyt, Richard, 4, 6, 84, 92, 163; *Principal Navigations* (1598), 92
Halley, Edmond, 80, 96–98, 99, 100, 101, 105, 118n85, 118n88, 134, 167, 200
Hatton, Christopher, 5, 21, 86
Henri II, king of France, 9, 83; Medici, Marie de, 13
Henri III, king of France, 10
Henri IV, king of France, 11–13, 14, 15, 18, 21, 24, 32, 35, 196; Medici, Catherine de, 9, 10. *See also* Henri of Navarre
Henri of Navarre, 10, 13, 15
Henry VIII, king of England, 5, 37, 82, 83, 92, 114n12
Herbert, William, 176, 199
Holland, Samuel, 152
Hollar, Wenceslaus, 27; frontispiece of Ogilby's *Britannia* (1675), 27
Hondius, Jodocus, 6
Hood, Thomas, 84, 92
Hooke, Robert, 27, 133
House of Commons, 24, 31, 37, 92, 135
hydra of heresy, 13, 26, 196
hydrographe de la marine (hydrographer of the navy), 90
hydrographic(al), xvi, xviii, xx, 80, 87, 88, 89, 90, 91, 93, 94, 96, 98, 100, 101, 102, 109, 110, 112, 113, 124, 137, 138, 139, 152, 153, 161, 162, 164, 165, 167, 169, 170, 171, 173, 175, 181, 183, 185, 186, 187, 199, 200
Hydrographic Office of the Navy. *See* *Dépôt des Cartes*
hydrography, xii, 84, 88, 92, 99, 113, 120n148, 182, 184, 188

Iberians, 84, 85, 86, 87, 162, 163
Iberville, Pierre Le Moyne, Sieur d', 127, 128, 130
Imperial Machine (Britain), 114, 152, 185, 187, 201
Imprimerie Royale, 90, 95
India House, 165, 169, 183, 184, 189n19
Interregnum, 91, 163

Jaillot, Alexis Hubert, 31, 32, 33, 35, 51, 53, 54; *Atlas Français* (1695), 32, 33, 35; *Atlas Nouveau* (1681), 31
James I, king of England and James VI of Scotland, 8, 23, 24–25, 26; Anne of Denmark, Queen consort (wife of James I), 8
James II, king of England, 30, 35, 37, 91, 93–94
Jefferys, Thomas, 45, 67–68, 77n127, 80, 103, 104, 106, 107–108, 124, 137, 138, 145, 146, 151, 152, 177–178, 196
Journal des Scavans, 33, 35

Kennebec River, 123, 142
Kent, county of, 47; King, Gregory, 47; Sheldon, Gilbert, Archbishop of Canterbury, 47, 48. *See also* Ogilby, John
Kersaint, Guy-Pierre de, 186
Kitchin, Thomas, 68, 103, 179

La Caille, Nicolas Louis de, 173, 174
La Favollière, 89, 90, 116n45
La Galissonière, Roland-Michel Barrin, marquis de, 138, 139, 173
La Hire, Philippe de, 49
La Houve, Paul de, 13
Languedoc, province of, 50, 51, 52, 55, 56, 64, 73n36; atlas project, 55, 64, 197; Estates General, 52, 55, 61, 73n35
La Pérouse, Jean-François de Galaup de, 185, 186
La Salle, René Robert Cavelier, Sieur de, 127, 129, 153n15
Lea, Philip, 58, 96
Le Clerc, Jean, 13
Leicestershire, county of, 69; Hastings, Francis, 69; Prior, John, 69
Le Maire, Père François, 129, 130, 198
Le Sueur, Pierre-Charles, 127, 128
letters of marque, 102
Linschoten, Jan Huygen van, 168
Lisieux, diocese of, 61, 62, 63; Brancas, Henri Ignace de, bishop of Lisieux, 61, 62. *See also* Anville, Jean Baptiste Bourguignon d'
livery companies of London, 22, 31
London Magazine, xix, 103, 105, 106, 107, 126, 178, 179
longitude, problem of, 80, 96, 100, 109
Lorient, 99, 118n96, 165, 172, 179, 180, 189n19, 192n101
Louis XIII, king of France, 18, 21, 31, 48
Louis XIV, king of France, xi, xvi, 2, 17, 18, 19, 20, 26, 31, 32, 33, 35, 37, 38, 43n120, 49, 53, 87, 89, 90, 93, 94, 95, 127, 129, 164, 168, 196; Anne of Austria (mother of Louis XIV), 17, 18; department of glory, 4, 31, 94; fabrication of image, xvi, 31; Sun King, 2, 4, 17, 21, 31, 32, 35, 37
Louis XV, king of France, 64, 65, 98, 102, 136, 144, 172, 197
Louis XVI, king of France, 109, 186
Louisiana, xix, 125, 126, 127, 129, 130–131, 134, 136, 154n26, 198
Lubin, Père Augustin, 48, 61

Machault, Jean-Baptiste de, 102, 119n116
Madras, 166, 169, 178, 180; Fort St. George, 162, 178, 180
magnetic variation, 96, 97, 99, 118n85
map consciousness, 5, 9
map making, x, xi, xii, xiii, xx, 27, 55, 76n111, 87, 91, 125, 145, 195, 200
maps, deictic function, 1, 19; instrument of power, xii; mimetic representation, 1, 19, 20; rhetorical texts, xiii; scientific discourse, xvii, 46, 49; tools of government, 5
marine chronometer, 80, 109, 187
Mascarenes, islands, 170, 173, 174, 182
Maurepas, Jean-Frédéric Phélypeaux, comte de, 98, 99, 104, 137, 142, 170. *See also Dépôt des Cartes*
Maurice of Orange, Prince, 86
Mazarin, cardinal Giulio, 17
Mead, Braddock, 145
Meaux, diocese of, 53, 60; Bossuet, Jacques Benigne (bishop of Meaux), 53, 60. *See also* Chevalier, François; Jaillot, Alexis Hubert
Medina, Pedro de, 83; *L'Art de Naviguer* (1554), 83
Mediterranean, 73n35, 83, 86, 87, 93, 95, 99, 104, 107, 108; chart of, 104; Levantine fleet, 87
Mercator, Gerhard, 10
Mercator projection (maps), 102, 103, 112, 167, 172
Middlesex, county of, 47, 58; Wolstenholme, Sir Thomas, 47, 48. *See also* Morgan, William; Ogilby, John
middling sort, 9, 91, 116n54, 178
Mignard, Pierre, *Louis at Maastricht* (1674), 32
Minorca, 104; battle of (1756), 104
Mississippi River, 127, 128, 129, 130, 133, 135, 136, 148, 150, 198

Mitchell, John, xix, 147, 148, 149, 150, 151, 198; *A Map of the British and French Dominions in North America* (1755), 147, 148, 149, 150, 151
Moll, Herman, 45, 60, 124, 132, 133, 149, 159n123
monsoons, 166, 168, 180, 185
Moore, John Hamilton, 111, 112, 113
Morgan, William, 27, 28, 30–31, 47, 72n15, 195
Mount, Richard, 58, 96; Mount and Page, xviii, 96, 98, 100, 101, 105, 111, 118n92, 161, 167, 170, 174, 176, 189n23, 199
Moxon, Joseph, 91; *A Book of Sea-plats* (1657), 91

Nairne, Thomas, 127, 128
national mapping project (France), xi, xvii, 46, 49, 52, 63, 66; second national (mapping) survey, xv, 56, 63, 65, 66
national marine atlas, xiv
navigation, 80–81, 82, 83, 84, 92, 111, 114n12, 140, 142, 147, 161, 166, 170, 172–173, 175, 177, 185, 187; challenges, xix, 103; instruments/tools, 80, 120n148; safety, xvi, xviii, 80, 92, 97, 99, 100, 113, 162, 169, 170, 171, 176, 180, 181, 184, 187, 199; works, xviii, 81, 83, 85, 88, 89, 92, 102, 103, 110, 114n5, 115n21, 137, 164, 165, 167, 169, 171, 174
Le Neptune François (1693), xviii, 79, 90, 93, 95–96, 107, 108
Neptune Oriental (1745-1775), 172, 175, 176, 179, 182, 192n104, 199
New France, 127, 128, 129, 137, 142, 143, 146, 147
Nicolay. *See* Vassallieu, Benedit de
Nicolay, Nicolas de, 9, 83. *See also* Medina, Pedro de
Nine Years' War, 33, 93, 104, 123, 124; King William's War, 123; Treaty of Ryswick (1697), 123. *See also* War of the League of Augsburg
Nismes, diocese of, 51, 73n35; Fléchier, Esprit, bishop of Nismes, 50, 51, 72n34. *See also* Nolin, Jean Baptiste
Noble, Charles, 161, 176
Nolin, Jean Baptiste, 33, 35, 51, 52, 53, 54, 71n3, 72n34, 73n48, 196
Norden, John, 21, 22, 23, 25, 41n74, 47, 59; *Civitas Londini* (1600), 23, 24; *A Description of the Honor of Windesor* (1607), 23; *A Guide for Cuntrey Men* (1623), 25; *Speculum Britanniae* or *Looking Glass of Britain*, 21, 23
Nouvelle France, 137, 142, 143, 144, 147. *See also* Canada; New France
Nova Scotia (Nouvelle Ecosse), xix, 123, 136, 137, 139, 140, 141, 142, 143, 144, 145, 146, 147, 148, 149, 150, 151, 157n96, 158n109, 159n123, 198. *See also* Acadie

Observatoire de Paris (Paris Observatory), xi, xii, xvi, 45, 49, 53, 70, 93
Ogilby, John, xvii, 25, 26–27, 28, 30, 31, 42n88, 42n96, 47–48, 49, 68, 72n15, 92, 195; *Britannia* (1675), 27, 30, 92
Order of the Garter, 8
Order of the Holy Spirit, 11, 15
Order of Saint Michael, 11, 15
Ordnance Survey (Board of Ordnance), xv, xviii, 46, 66, 70, 71, 197, 198, 200
Orléans family, 61; Gaston d'Orléans, uncle of Louis XIV, 17; Philippe d'Orléans, regent for Louis XV, 55
Orry, Philibert, 63, 65
Ortelius, Abraham, 10; *Theatrum Orbis Terrarum* (1570), 5, 10

Paris, diocese of, 53; Harlay, François de, archbishop of Paris, 53; Jouvin de Rochefort, Albert, 53
Paris, généralité, 53; Charron, Jean Jacques (intendant of Paris), 53. *See also* Jaillot, Alexis Hubert
Parliament, xvi, 2, 4, 24, 26, 28, 30, 35, 37, 40n23, 46, 65, 76n121, 92, 93, 135, 136, 178, 195
Pène, Charles, 80, 90, 96, 116n48, 116n49
Pepys, Samuel, 91–93, 95, 95–96, 98, 117n66
Philip II, king of Spain (and the Netherlands), 5, 85
Philippine Islands, 173, 179, 180
Picard, Jean, 49, 52

Pickles, John, xiii, 125, 198
Pigneau de Béhaine, Pierre-Joseph-Georges, 186; Nguyen Anh (Gia Long), 185, 186
Pitt's Passage, 184; Wilson, William, 180
Plantin, Christophe, 85. *See also* Waghenaer, Lucas Janszoon
Plessis, Armand du, Cardinal Richelieu, 48, 87
Plymouth, 58, 67, 97, 99, 101
Ponant (Atlantic fleet), 87, 90
Pontchartrain, Jérôme Phélypeaux, comte de, 118n80, 119n115, 128
Pontchartrain, Louis Phélypeaux, comte de, 95, 129
Ponts et Chausées (Bridges and Roads), 63
Popple, Alured, 128, 143
Popple, Henry, 124, 134, 143, 144, 155n44, 159n123; *A Map of the British Empire in America* (1733), 134, 137, 142–143, 144, 150
Portuguese, xx, 84, 114n5, 163, 168, 172, 173, 178, 182
premier géographe du roi (first royal geographer), 45, 54
pretensions (territorial). *See* encroachments
privateers, 89, 93, 95, 101, 102

Quebec, 128, 130, 136, 138, 142
Quiberon Bay, battle of (1759), 99, 106

Raleigh, Sir Walter, 6
Raudot, Antoine-Denis, 129, 150
Reims, diocese of, 48; Jubrien, Jean, 48, 61; Sainte-Marie, Gabriel de (archbishop of Reims), 48
Restoration, 26, 27, 31, 91, 92, 163
Reunions, 35
rhetoric, xiii, xv, xix, 1, 2, 3–4, 18, 24, 26, 28, 30, 35, 52, 82, 139, 147, 151, 198; power, xvii, 38, 125, 145, 148, 152; scientificity, 38, 80, 87, 148; science, 96
Richelieu, cardinal. *See* Plessis, Armand du
Richelieu, Emmanuel-Armand de Vignerot du Plessis de, duc d'Aiguillon, 103, 119n115
Rochefort, 104, 107
Rosily-Mesros, François-Etienne de, 185, 186
Rotz, Jean, 82; *Boke of Idrography* (1542), 82
Rouillé, Antoine-Louis, comte de Jouy, 99, 102; *See also Dépôt des Cartes*
Roy, General William, 68, 70. *See also* Ordnance Survey
Royal Hospital School (Greenwich), 92
Royal Society (London), xvi, 27, 46, 58, 59, 66, 69, 70, 80, 96, 97, 98, 101, 118n85, 124, 134, 188n3, 200; Banks, Joseph (president), 70, 188n3

sailing directions, xiv, 80–81, 83, 84–85, 86, 96, 102, 103, 108, 114n5, 115n21, 161, 164, 166–168, 169, 170, 179, 183, 191n61
Sanson, Nicolas, 31, 48–49, 50, 126; Sansons (Guillaume and Adrien, sons of Nicolas), 32
Sauveur, Joseph, 90, 96
Saxton, Christopher, 3, 5, 6, 8, 9, 21, 23, 24, 40n23, 47, 48, 59, 195; *Atlas of the Counties of England and Wales* (1579), 5, 47, 195
Sayer, Robert (1725-1794), 110, 111, 111–113, 121n154, 184. *See also* Bennett, John
scientific authority, 80, 82, 87, 124, 149, 152, 170, 198; discourse, xiii, xvii, 38, 46, 52, 98, 170, 174, 199; expeditions, 46, 80, 109, 186, 188, 199, 200; geography, 45, 198; knowledge, 80, 130, 148, 162, 171; methodology, 54, 125, 153; scientificity, 38, 80, 87, 99, 148, 185
Seckford, Thomas, 5, 6
Seguin, Joseph-Dominique, 64, 65, 67, 71, 197
Seignelay, marquis de. *See* Colbert, Jean-Baptiste Antoine, marquis de Seignelay
Seller, John, 91, 92, 100, 101, 105, 117n66, 165, 166
Seneffe, battle of, 32
Senex, John, 45, 58, 59
Senex, Mary, 101
Senlis, diocese of, 54, 55; Chamillard, Jean François de, bishop of Senlis, 54. *See*

also Delisle, Guillaume
Seven Years' War (1756-1763), 65, 68, 101, 105, 106, 108, 109, 125, 130, 136, 151, 159n129, 161, 169, 179, 180, 192n101, 198
Society for the Encouragement of Arts, Manufactures and Commerce, xviii, 66, 67, 68, 69, 196, 197; Baker, Henry, 66; Borlase, William, 66, 71; Rawthmill Coffee House (Covent Gardens), 66
Society of Sciences at Montpellier, 56, 64, 73n35, 197
Soissons, généralité of, 63; Méliand, Charles-Blaise, 63. *See also* Anville, Jean Baptiste Bourguignon d'
South Sea Bubble affair, 132, 155n43
Spanish Armada (1588), 6, 8, 84
Speed, John, 8, 47; *Theater* (1611), 8
Spice Islands, 81, 162, 163, 177
St. Cast, battle of, 102, 105, 107, 120n124
St. Lawrence River, 123, 133, 140, 142, 143, 144
St. Malo, 104, 107, 120n123
Stavorinus, Captain J. S., 177, 185
strategic axis, 128, 130, 136
Sulu Sea, 180; Sultan of Sulu, 181
Sunda Straits, 181, 184, 185
Surrey, county of, 58, 59. *See also* Senex, John
Sussex, county of, 70, 71; Budgen, Richard, 75n74; Gardner, William, 70, 71; Yeakell, Thomas, 70. *See also* Senex, John

Thames School, chart making, 57, 58, 165, 172
Thirty Years' War, 17, 24, 163
Thornton, John, 58, 165–167, 170, 173, 174, 175, 176, 189n30, 199
Toulon, 95, 104, 107; battle of, 99
Tassin, Christophe, 87
Tourville, Anne Hilarion de Constantin, comte de, 94, 95
Toulouse, diocese of, 50; Colbert de Saint Pouange, Jean Baptiste Michel (archbishop of Toulouse), 50; Gautier, Hubert, 50. *See also* Jaillot, Alexis Hubert
Treaty of Aix la Chapelle (1748), 99, 138
Treaty of Dover (1670), 89
Treaty of Paris (1763), 152, 199
Treaty of Utrecht (1713), 128, 129, 137, 138, 142, 145, 147, 155n65
triangulation, xiv, 45, 49, 51, 56, 59, 60, 64, 65, 67, 69, 70, 71, 79, 93, 117n70, 128, 147, 152, 173, 196
trigonometric survey. *See* triangulation
Trinity House, 83, 92, 96, 110, 114n12, 117n66; Borough, Stephen, 83, 114n12. *See also* Deptford Strand
Trudaine, Daniel, 65
Turnpike Trusts, 66

United Netherlands Chartered East India Company (VOC). *See* Dutch East India Company

Van Keulen, Johannes, 165, 167, 174, 183
Vassallieu, Benedit de (dit Nicolay), 13, 14, 15, 19, 21
Vaugondy, Didier Robert de, 144, 145, 147, 148, 158n104, 158n106
Vivier, David du, 52, 53

Waghenaer, Lucas Janszoon, 85–86, 91, 104, 108, 110; *Enchuyser Zeecaertboeck* (1598), 86; *Speculum Nauticum* (1586), 86; *Spieghel der Zeevaerdt* (1584), 85, 86; *Thresoor der Zeevaert* (1592), 86. *See also* Ashley, Anthony
War of American Independence, 70, 110, 152, 179, 184
War of Austrian Succession, 99, 136, 138, 156n71, 169
War of the League of Augsburg, 33, 43n120, 93
War of Spanish Succession, 60, 129
Wars of Religion (France), 9, 10, 13, 83
Warwickshire, county of, 59, 61. *See also* Beighton, Henry
Western Approaches, 99, 106
West Indies, 99, 105, 108, 138, 163, 168, 177
Whiston, William (1657-1752), 100–101. *See also* Board of Longitude
Whitehall, xvi, 31, 131

William of Orange (William III, king of England), 37, 85, 93–94, 123, 127, 166; William and Mary, 37, 92, 94
Woutneel, Hans, 8

About the Author

Christine Marie Petto earned her BA in astronomy (1984) from Boston University, and her MA in history of science (1992) and PhD in early modern European history (1996) from Indiana University, Bloomington. Currently, she is a professor of early modern European history and an assistant to the Deans of Arts and Sciences at Southern Connecticut State University. She is also a coeditor of a transdisciplinary and interdisciplinary journal, *Environment, Space, Place*, committed to the presentation of works that address the "geographical turn" or the spatial aspects of worldly phenomena. Her area of research is English and French map making in the early modern period and the relationship between map production and governmental power. Her first monograph, *When France was King of Cartography* (Lexington Books, 2007), addressed the role of map production and governmental support in the seventeenth and eighteenth centuries. In addition to her work on power and map making, she has published on women in map making for *Cartographica* 44, no. 2 (2009), "Playing the Feminine Card: Women of the Early Modern Map Trade," as well as contributing an entry, "Women and Cartography," for volume four of the History of Cartography Project (forthcoming). Her recent projects range from the post roads maps of England and France to an investigation of the relationship between the medieval ritual of "beating the bounds" and early modern mapping to pirates and privateers in the early modern charting of the Mediterranean Sea.